ETO MULTICENTER

MOLECULAR INTEGRALS

ETO Multicenter Molecular Integrals

Proceedings of the First International Conference held at Florida A&M University, Tallahassee, Florida, U.S.A., August 3-6, 1981

edited by

Charles A. Weatherford
Institute for Molecular Computations, Florida A&M University, Tallahassee, Florida, U.S.A.

and

Herbert W. Jones
Physics Department, Florida A&M University, Tallahassee, Florida, U.S.A.

D. Reidel Publishing Company
Dordrecht, Holland / Boston, U.S.A. / London, England

Library of Congress Cataloging in Publication Data

International conference on ETO multicenter molecular integrals (1st : 1981 : Florida Agricultural and Mechanical University)
ETO multicenter molecular integrals.

Includes indexes.
1. Molecular orbitals–Congresses. 2. Quantum chemistry–Congresses. I. Weatherford, C. A. (Charles A.), 1947– II. Jones, Herbert W. III. Title. IV. Title: ETO multicenter molecular integrals. V. Title: Molecular integrals.
QD461.I58 1981 541.2'8 82-18105

ISBN-13: 978-94-009-7923-9 e-ISBN-13: 978-94-009-7921-5
DOI: 10.1007/978-94-009-7921-5

Published by D. Reidel Publishing Company
P.O. Box 17, 3300 AA Dordrecht, Holland

Sold and distributed in the U.S.A. and Canada
by Kluwer Boston Inc.,
190 Old Derby Street, Hingham, MA 02043, U.S.A.

In all other countries, sold and distributed
by Kluwer Academic Publishers Group,
P.O. Box 322, 3300 AH Dordrecht, Holland

D. Reidel Publishing Company is a member of the Kluwer Group

CONTENTS

PREFACE

The First International Conference on ETO Multicenter Molecular Integrals was held August 3-6, 1981, on the Florida A&M University campus in Tallahassee, Florida, USA. Thirty four scientists from eight countries assembled in Tallahassee under the sponsorship of the Institute for Molecular Computations and the Physics Department at Florida A&M. Financial support is gratefully acknowledged from the National Science Foundation, U.S. Army Research Office (Durham), Office of Naval Research, the National Aeronautics and Space Administration (NASA), and Florida A&M University. In particular, the editors would like to thank Dr. Joe Majowicz and Dr. David Squire of the U.S. Army, and Dr. Aaron Temkin of NASA for their support and encouragement.

We would also like to acknowledge the Atlanta University Resource Center for Science and Engineering for financial support in the preparation of the manuscript. Also, of course, we sincerely appreciate the participation of the attendees and especially the contributors to this work. As a result of their presentations, the conference was a very intense and fertile forum for the exchange of ideas on a very important and historic problem of quantum chemistry.

Finally, we want to thank Ms. Sonja Richardson for the enthusiastic, diligent and competent preparation of a very difficult manuscript.

Charles A. Weatherford Herbert W. Jones

C. A. Weatherford and H. W. Jones (eds.), ETO Multicenter Molecular Integrals, vii.

Dedicated to Per-Olov Löwdin, who personally worked on molecular integrals and encouraged others to do so.

PARTICIPANTS

Danko Antolovic
Department of Chemistry
The Johns Hopkins University
Baltimore
Maryland 21281 USA

N.A. Baki Baykara
Marmara Scientific and Industrial Research Institute
Applied Mathematics Department
P.O. Box 21 Gebze
Kocaeli
Turkey

Ante Graovac
Institut für Strahlenchemie
Max Planck Institut
Stift Strassee 34-36
D-4330 Mulheim a.d. Ruhr 1
West Germany

Peter Habitz
IBM Corporation
P.O. Box 390
Poughkeepsie
New York 12602 USA

Stan Hagstrom
Department of Chemistry
University of Indiana
Bloomington
Indiana 47405 USA

John Hall
Department of Chemistry
Morehouse College
Atlanta
Georgia 30314 USA

Frank E. Harris
Department of Physics
University of Utah
Salt Lake City
Utah 84112 USA

Babu L Jain
Institute for Molecular Computations
and
Physics Department
Florida A&M University
Tallahassee
Florida 32307 USA

Herbert W. Jones
Institute for Molecular Computations
and
Physics Department
Florida A&M University
Tallahassee
Florida 32307 USA

Chew-Lean Lee
Box 981
Physics Department
Florida A&M University
Tallahassee
Florida 32307 USA

Ker-Fong Lee
Department of Chemistry
Florida State University
Tallahassee
Florida 32306 USA

Otieno Malo
Department of Physics
University of Nairobi
Box 30197
Nairobi
Kenya

Harvey Michels
Department of Physics
United Technologies Research Center
East Hartford
Connecticut 06108 USA

Hank Monhorst
Department of Physics
University of Florida
Gainesville
Florida 32611 USA

Alfred Msezane
Department of Physics
Morehouse College
Atlanta
Georgia 30314 USA

Harry Partridge
NASA Ames Research Center
Mail Stop 230-3
Moffett Field
California 94035 USA

Russell M. Pitzer
Department of Chemistry
Ohio State University
140 W. 18th Avenue
Columbus
Ohio 43210 USA

Milan Randic
Ames Laboratory, US-DOE
Iowa State University
Ames
Iowa 50011 USA

M.A. Rashid
International Centre for Theoretical Physics
Trieste
Italy
and
Department of Mathematics
Ahmadu Bello University
Zaria
Nigeria

C.C.J. Roothaan
Department of Chemistry
University of Chicago
Chicago
Illinois 60637 USA

Klaus Ruedenberg
Departments of Chemistry and Physics
Iowa State University
Ames
Iowa 50010 USA

R.R. Sharma
Department of Physics
University of Illinois
Chicago
Illinois 60680 USA

Isaiah Shavitt
Battelle Memorial Institute
505 King Avenue
Columbus
Ohio 43201 USA

David M. Silver
Applied Physics Laboratory
The Johns Hopkins University
Laurel
Maryland 20810 USA

Harris J. Silverstone
Department of Chemistry
The Johns Hopkins University
Baltimore
Maryland 21218 USA

E. Otto Steinborn
Institut für Physikalische and Theoretische Chemie
Universität Regensburg
D-8400 Regensburg
West Germany

James D. Talman
Department of Applied Mathematics and Physics
and
Centre for Chemical Physics
University of Western Ontario
London
Ontario, Canada N6A 5B9

William P. Tucker
Physics Department
Florida A&M University
Tallahassee
Florida 32307 USA

Ralph W. Turner
Department of Chemistry
Florida A&M University
Tallahassee
Florida 32307 USA

Arun S. Wagh
Department of Physics
University of the West Indies
Kingston 7
Jamaica
West Indies

Arnold C. Wahl
Science Applications, Inc.
1211 West 22nd Street
Oakbrook
Illinois 60521 USA

Charles A. Weatherford
Institute for Molecular
Computations
and
Physics Department
Florida A&M University
Tallahassee
Florida 32307 USA

S. Wozniak
Department of Chemistry
Florida State University
Tallahassee
Florida 32306 USA

Bronislaw Zurawski
Computing Centre
Nicholas Copernicus University
87-100 Torun, ul.
Gagarina 7
Poland

PARTICIPANT CONTRIBUTORS

Peter Habitz
IBM Corporation
P.O. Box 390
Poughkeepsie
New York 12602
USA

Frank E. Harris
Department of Physics
University of Utah
Salt Lake City
Utah 84112
USA

Babu L. Jain
Institute for Molecular Computations
and Physics Department
Florida A&M University
Tallahassee
Florida 32307
USA

Herbert W. Jones
Institute for Molecular Computations
and
Physics Department
Florida A&M University
Tallahassee
Florida 32307
USA

Ker-Fong Lee
Department of Chemistry
Florida State University
Tallahassee
Florida 32306
USA

Harvey Michels
Department of Physics
United Technologies Research Center
East Hartford
Connecticut 06108
USA

Russell M. Pitzer
Department of Chemistry
Ohio State University
140 W. 18th Avenue
Columbus
Ohio 43210
USA

Milan Randic
Ames Laboratory, US-DOE
Iowa State University
Ames
Iowa 50011
USA

M.A. Rashid
International Centre for Theoretical Physics
Trieste
Italy
and
Department of Mathematics
Ahmadu Bello University
Zaria
Nigeria

R.R. Sharma
Department of Physics
University of Chicago
Chicago
Illinois 60680
USA

Harris J. Silverstone
Department of Chemistry
The Johns Hopkins University
Baltimore
Maryland 21218
USA

E. Otto Steinborn
Institut für Physikalische
and Theoretische Chemie
Universität Regensburg
D-8400 Regensburg
West Germany

James D. Talman
Department of Applied Mathematics
and Physics
and
Centre for Chemical Physics
University of Western Ontario
London, Ontario
Canada N6A 5B9

Charles A. Weatherford
Institute for Molecular
Computations
and Physics Department
Florida A&M University
Florida 32307
USA

NON-PARTICIPANT CONTRIBUTORS

Enrico Clementi
IBM Corporation
P.O. Box 390
Poughkeepsie
New York 12602
USA

Robert D. Cloney, S.J.
Department of Chemistry
Fordham University
Bronx
New York 10458
USA

Kenneth C. Daiker
CONDUIT
Box 388
Iowa City
Iowa 52244
USA

I.I. Guseinov
Physical Faculty
Azerbaijan State University
23 Lumumba Street
Baku
USSR

Richard K. Moats
BDM Corporation
7915 Jones Branch Drive
McLean
Virginia 22102
USA

H. David Todd
Department of Chemistry
Wesleyan University
Middletown
Connecticut 06457
USA

Conference Participants (left to right):
1st row; Hall, Harris, Shavitt, Jain, Jones, Weatherford, Rashid, Malo, Roothaan, Sharma, Tucker
2nd row; Randić, K.F. Lee, C.L. Lee, Msezane, Talman, Graovac, Silverstone, Ruedenberg, Antolovic, Steinborn
3rd row; Wozniak, Wagh, Partridge, Hagstrom, Zurawski, Monkhorst, Habitz, Baykara, Silver, Michels, Pitzer.

INTRODUCTION

In recent years, the literature in the fields of theoretical molecular physics and ab-initio quantum chemistry has evidenced a renewed interest in obtaining accurate, efficient algorithms for performing multicenter integrals, which result from the use of exponential functions as basis sets for LCAO-MO calculations. This is a problem as old as quantum chemistry itself. Indeed, "the problem" may be considered a classic problem of quantum chemistry.

Gaussian orbitals were introduced in the 50's and 60's to partially circumvent difficulties; however, because exponential orbitals are universally recognized as superior functions, and because many more Gaussians than exponentials are required for a given level of accuracy, the problem has continued to remain an active area of research.

Inspite of its importance, there has never been a conference devoted solely to consideration of multicenter integrals over exponential functions. The Proceedings of the First International Conference on ETO Multicenter Molecular Integrals is the result of a conference devoted to this one problem, and describes some of the more recent ideas invented for the solution. Included are the latest developments in the use of Fourier Transforms and Spherical Bessel Transforms, expansions in complete, orthonormal ETO's, techniques of computer algebra, and Löwdin alpha function methods.

It is the hope of the editors that the proceedings of the conference will have the desired result of bringing ever-larger molecular systems within the realm of LCAO-MO methods by aiding in the synthesis of more efficient and accurate algorithms for integrals computation. Consistent with this desire, we would like to note here that one paper (Habitz and Clementi), although not directly concerned with the ETO problem, generally discusses strategies for treating large molecular systems by the systematic neglect of integrals that are shown to have a small effect.

In addition, two papers (Ker-Fong Lee and Guseinov) which were not actually presented at the conference, but were contributed to the proceedings, are collected at the end of the manuscript.

Charles A. Weatherford   Herbert W. Jones

C. A. Weatherford and H. W. Jones (eds.), ETO Multicenter Molecular Integrals, 1.

EXPERIENCE USING SPHERICAL HARMONIC EXPANSIONS TO EVALUATE MOLECULAR INTEGRALS

Russell M. Pitzer
Department of Chemistry, Ohio State University, 140 W. 18th Avenue, Columbus, Ohio 43210

It is clear from the presentations made at this conference by Harris, Silverstone, Sharma, Jones, Michels, Rashid, Steinborn, Weatherford, Antolovic, and Silver, using quantities variously labeled α, β, B, C, v, ζ, that there is continuing interest in evaluating multicenter Slater (exponential) orbital molecular integrals by the spherical harmonic expansion method, or by equivalent methods which give the same form to the final result. The equivalence of various formalisms of this type is easily recognized for 3-center integrals involving all 1s orbitals. If the final formula depends on the geometrical parameters of the three centers (2 internuclear distances and one angle) in terms of an infinite series of Legendre polynomials of the cosine of the angle, then the formula is equivalent to the one obtained by using spherical harmonic expansions about one nucleus.

We used this method in the context of the zeta-function formalism[1] from approximately 1960 to 1972, and the purpose of this paper is to comment on our experience with it. When I first became involved in this work in 1960, there were already several people in the Solid State and Molecular Theory Group (J. C. Slater, Director) at M.I.T. writing molecular integral programs[2] using the zeta-function method; M. P. Barnett, L. A. Burnelle, D. E. Ellis, D. P. Merrifield, and J. W. Moskowitz. During the next few years, C. A. Christy, C. W. Kern, W. E. Palke, B. Woznick, and J. P. Wright also participated in this work; some of the latter were in the W. N. Lipscomb group at Harvard. An extensive comparison of integral values was carried out in 1962 in correspondence with M. Karplus and I. Shavitt at Columbia University and Watson Laboratory, who were simultaneously developing the Gaussian-transform method.[3] Integral values obtained with these and later versions of the zeta-function programs were published for some molecules.[4,5]

When the zeta-function programs for 3- and 4-center integrals involving s and p orbitals were moderately refined, they were submitted to the Quantum Chemistry Program Exchange.[6] They were written

C. A. Weatherford and H. W. Jones (eds.), ETO Multicenter Molecular Integrals, 3–6.

in FORTRAN II and used single precision (36 bits) on the IBM 709-7090-7094 series computers. The zeta functions were assembled from spherical Bessel functions with the extensive use of recurrence relations.[1] The associated Legendre functions were also obtained using recurrence relations. The radial integrals were evaluated using single or double Gaussian quadratures with Gauss-Legendre and Gauss-Laguerre points. Although set up to do individual integrals, the programs reused intermediate data in various amounts, sometimes even more than is implied by a charge distribution approach, so that they could be quite efficient if integrals were requested in the correct order. The first extensive application of these programs was in computing the barrier to internal rotation in ethane.[4,7]

Later versions of the zeta-function programs, generalized to include integrals with d orbitals, were chained together to make complete calculations much more practical.[8] These were used until the IBM 7094 computers were phased out in the early 1970's. At that time a major programming effort would have been necessary to continue the use of the programs. In order to make this worthwhile, considerable improvements would have been necessary if the programs were to be competitive with (1) Gaussian-transform programs[3,9], and (2) contracted Gaussian programs.[10] It seemed to me that important improvements were possible, particularly in the numerical quadratures, so that (1) might be accomplished for triatomic molecules, at least, but (2) seemed unlikely. Perhaps some of the approaches described at this conference will alter these comparisons.

Aside from relative computational efficiencies, the difficulty most often mentioned with respect to spherical harmonic expansion methods is the convergence of the resulting series. To describe this problem in more detail, let us use 3-center, all 1s integrals as examples. The three centers are labeled A, B, and C, and A is the expansion center. The resulting series have terms labeled by the ℓ index of the Legendre functions of the angle θ_{BAC}. The radial integrals depend on the orbital exponents and on the distances R_{AB} and R_{AC}. LaBudde and Sahni[11] analyzed the convergence of these series and concluded that, asymtotically, the series are all geometric in the ratio of the smaller of the two distances to the larger. There are also various inverse powers of ℓ for various cases which help the convergence, especially in the case of equal distances. We observed this geometric convergence numerically in a number of cases which were examined in detail. The effect of large orbital exponents on A seemed to be to increase the convergence of the first few terms while large orbital exponents on B or C had the opposite effect.

Our programs were usually dimensioned for 30 to 40 terms, and, to the accuracy standard used (usually 10^{-6} h), there were very few cases of insufficient convergence. Our recourses in such cases were to recompute the integral either with more terms or by using a different expansion center. Of the three types of 3-center integrals which appear in the usual energy expression, the nuclear attraction and hybrid (coulomb) integrals could be expanded about any one of the three centers, while for the exchange type only one center could be used without treating the integral essentially as a 4-center integral. Fortunately, for a given set of parameters, the exchange integrals generally converged the fastest. A particular case of slow convergence about all three centers was found for an electric field gradient integral in formaldehyde.[12] A promising development with respect to the convergence of these series was the demonstration[13] that nonlinear transformations of the terms significantly accelerated the convergence.

It was interesting to hear, at this conference, about continuing efforts to make the use of Slater (exponential) orbitals more competitive in molecular theory. I think that most of the workers in this field would be delighted to see these efforts succeed.

REFERENCES

1. Barnett, M.P., Coulson, C.A.: 1951, Phil. Trans. Roy. Soc. 243, p.221. Barnett, M.P.: 1963, Meth. Comp. Phys. 2, p.95.

2. Many notes concerning this effort appeared in the Quarterly Progress Reports of the Solid State and Molecular Theory Group.

3. Shavitt, I. and Karplus, M.: 1962, J. Chem. Phys. 36, p.550. Karplus, M. and Shavitt, I.: 1962, J. Chem. Phys. 38, p.1256. Shavitt, I.: 1963, Meth. Comp. Phys. 2, p.1. Shavitt, I. and Karplus, M.: 1964, J. Chem. Phys. 43, p.398. Kern, C.W. and Karplus, M.: 1964, J. Chem. Phys. 43, p.415.

4. C_2H_6: Pitzer, R.M.: 1963, thesis, Harvard University.

5. H_2O: Pitzer, R.M. and Merrifield, D.P.: 1970, J. Chem. Phys. 52, p.4782.

6. Pitzer, R.M., Wright, J.P. and Barnett, M.P.: 1964, QCPE 1, 22-25. QCPE 22,23 were changed to FORTRAN IV by Macias, A. and Leffler, L.: 1966, QCPE 3, 86,87.

7. Pitzer, R.M. and Lipscomb, W.N.: 1963, J. Chem. Phys. 39, p.1995.

8. Palke, W.E. and Lipscomb, W.N.: 1966, J. Amer. Chem. Soc. 88, p.2384.

9. Stevens, R.M.: 1970, J. Chem. Phys. 52, p.1397.

10. Neumann, D.B., Basch, H., Kornegay, R.L., Synder, L.C., Moskowitz, J.W., Hornback, C. and Liebman, S.P.: 1971, QCPE 8, 199. Dupuis, M., Rys, J. and King, H.: 1977, QCPE 14, 336. Binkley, J.S., Whitehead, R.A., Hariharan, P.C., Seeger, R. and Pople, J.A.: 1978, QCPE 15, 368.

11. LaBudde, C.D. and Sahni, R.C.: 1960, J. Chem. Phys. 33, p.1022.

12. Flygare, W.H., Pochan, J.M., Kerley, G.I., Caves, T., Karplus, M., Aung, S., Pitzer, R.M. and Chan, S.I.: 1966, J. Chem. Phys. 45, p.2793.

13. Petersson, G.A. and McKoy, V.: 1967, J. Chem. Phys. 46, p.4362.

ON THE EVALUATION OF ETO MOLECULAR INTEGRALS BY SERIES EXPANSIONS USING COMPLETE FUNCTION SETS

E. Otto Steinborn
Institus für Physikalische und Theoretische Chemie,
Universität Regensburg, D-8400 Regensburg, West Germany

1. INTRODUCTION

In quantum chemistry the common approach to construct the molecular wavefunction by molecular orbitals (MO's) which are linear combinations of atomic orbitals (AO's), i.e. the LCAO-MO-method, leads to the difficult matrix elements of the Hamiltonian operator which are the notorious molecular multicenter integrals. For exponential-type atomic orbitals (ETO's), these integrals are hard to evaluate. To by-pass this difficulty, the Gaussian-type orbitals (GTO's) were introduced.[1] Because of the great number of GTO's required in a molecular calculation, other problems occur if these orbitals are used. We want to discuss molecular integrals over ETO's. The various functions (AO's) as well as the operator in the integrand of each integral are defined with respect to different centers, i.e. the positions of the various atomic nuclei in a molecule, whereas in order to perform the integration, the integration variables are to be defined with respect to one common center. Therefore, it is necessary to express all those exponential-type functions (ETF's) and operators, which occur in the integrand, with respect to the chosen fixed reference system of integration. This causes serious problems because by doing so these functions and operators may transform to very complicated mathematical expressions, although they may be of relatively harmless mathematical form originally when each function (AO) is defined with respect to its "atomic coordinate system" centered at the position of the respective atomic nucleus.

Of course, a molecular integral is the harder to be evaluated the more different centers are needed to define the various origins of the "atomic coordinate systems" of the orbitals and operators in its integrand. One can try to simplify such an integral by reducing the number of centers used for its definition.

The simplest example for an integral with more than one center is the two-center-one-electron integral

C. A. Weatherford and H. W. Jones (eds.), ETO Multicenter Molecular Integrals, 7–27.

$$\Phi(\vec{R}) = \int d\tau\ g(\vec{r}_A)\ f(\vec{r}_B). \tag{1.1}$$

With respect to an arbitrary origin 0, the local vector $\vec{r}$ specifies the position of the elctron, whereas $\vec{A}$ and $\vec{B}$ specify the positions of the centers A and B, respectively. Then, g and f may represent AO's on the nuclei A and B, respectively, with $\vec{r}_A = \vec{r}-\vec{A}$ and $\vec{r}_B = \vec{r}-\vec{B}$; we shall also denote such functions by g_A and f_B. Without loss of generality the origin 0 may be chosen to coincide with A. Obviously, with given functions g and f, the *overlap or convolution integral* Eq. (1.1) becomes a function of the distance $\vec{R}$,

$$\Phi(\vec{R}) = \int d\vec{r}\ g(\vec{r})\ \ f(\vec{r}-\vec{R}). \tag{1.2}$$

One method for evaluating such and other multicenter integrals utilizes the Fourier transform convolution theorem

$$\bar{\Phi}(\vec{k}) = \bar{g}(\vec{k})\ \bar{f}(\vec{k})\ \ (2\pi)^{3/2} \tag{1.3}$$

for the Fourier transforms $\bar{\Phi}$, $\bar{g}$, and $\bar{f}$, respectively. With its help, the two-center overlap integral $\Phi(\vec{R})$ of Eq. (1.1) is transformed into the one-center Fourier integral

$$\begin{aligned}\Phi(\vec{R}) &= \int d\vec{k}\ \exp(i\vec{k}\cdot\vec{R})\ \bar{\Phi}(\vec{k})\ \ (2\pi)^{-3/2} \\ &= \int d\vec{k}\ \exp(i\vec{k}\cdot\vec{R})\ \bar{g}(\vec{k})\ \ \bar{f}(\vec{k})\end{aligned} \tag{1.4}$$

which sometimes may be easier to evaluate. However, this method requires the evaluation of the three integrals $\bar{f}$, $\bar{g}$, and $\Phi(\vec{R})$ of Eqs. (1.3,4) instead of one, and these integrals may sometimes also be hard to evaluate. Especially the back transformation may cause problems. Therefore, although the Fourier transform method has certainly its great advantages and merits, we chose to simplify the multicenter integrals by transformations in configuration space.

2. TWO-RANGE ADDITION THEOREMS AND BI- AND POLYPOLAR EXPANSIONS

The transformations considered in detail are especially those which make it possible to shift an AO or operator from one center to another. This can be achieved if an addition theorem for the function under consideration can be derived.

For a function f depending on the local vector $\vec{\rho} = \vec{r}_1 + \vec{r}_2$ an addition theorem does exist if it is possible to express $f(\vec{\rho})$ by a series

$$f(\vec{r}_1+\vec{r}_2) = \sum_\lambda \sum_\mu A_{\lambda\mu}\ U_\lambda\ (\vec{r}_1)\ V_\mu(\vec{r}_2) \tag{2.1}$$

in which the variables $\vec{r}_1$ and $\vec{r}_2$ are separated on the R.H.S. of the equation. Only for some classes of functions it is possible to derive such expansions with given functions U_λ and V_μ, for which the expansion coefficients $A_{\lambda\mu}$ are explicitly known (and can be programmed). Considering orbitals, it is advantageous to separate the angular variables from the radial variables by expanding furthermore

$$U_\lambda(\vec{r}_1) = \sum_{\ell,m} u_{\lambda\ell}(r_1) \; Y_\ell^m(\theta_1,\phi_1), \tag{2.2a}$$

$$V_\mu(\vec{r}_2) = \sum_{\ell,m} v_{\lambda\ell}(r_2) \; Y_\ell^m(\theta_2,\phi_2). \tag{2.2b}$$

Then, $f(\vec{r}_1+\vec{r}_2)$ is expressed by series expansions in terms of spherical harmonics and functions depending on either one of the radial variables r_1 and r_2. Further intrinsic symmetry properties of such addition theorems can advantageously be used for integral evaluations.[2,3]

It is clear that for $\vec{r}_1 = \vec{r}$ and $\vec{r}_2 = -\vec{R}$ the relation Eq. (2.1) expresses a function $f(\vec{r}-\vec{R})$, that is shifted by $\vec{R}$ from O, by functions $U(\vec{r})$ and $V(\vec{R})$ centered at O. If introduced into Eq. (1.2), the two-center integral becomes a sum of one-center integrals. Of course, an addition theorem is in fact a one-center expansion.

We derived such addition theorems firstly for simple functions[4,5] later for more complicated functions[6], including various ETF's of interest[7,8], especially Slater-type functions *(STF's, or orbitals STO's)*

$$\chi_{N,L}^{M}(\alpha\vec{r}) = e^{-\alpha r}(\alpha r)^{N-1}\cdot Y_L^M(\Omega_{\vec{r}}) \tag{2.3}$$

as well as B-functions

$$B_{N,L}^{M}(\alpha r) = \hat{k}_{N-1/2}(\alpha r)\mathcal{Y}_L^M(\alpha\vec{r}) \; [(2N+2L)!!]^{-1}, \tag{2.4a}$$

$$\hat{k}_{N-1/2}(r) = r^{-1} e^{-r} \sum_{p=1}^{N} \frac{(2N-p-1)!}{(p-1)!\,(N-p)!} \; 2^{p-N} r^p = (2/\pi)^{1/2} \, r^{N-1/2} K_{N-1/2}(r). \tag{2.4b}$$

Here, $\hat{k}_\nu$ is the so-called reduced Bessel function (RBF), originally introduced by Shavitt[9]. It is closely related to the modified Bessel function of the second kind K_ν.[10] The surface spherical harmonics $Y_\ell^m(\theta,\phi)$ are defined in Condon-Shortley phases[11]. The regular solid spherical harmonic $r^\ell Y_\ell^m(\theta,\phi)$ is written as $\mathcal{Y}_\ell^m(\vec{r})$ in Eq. (2.4a).

We use $\mathcal{Z}_\ell^m(\vec{r})$ for the irregular solid spherical harmonic $r^{-\ell-1}Y_\ell^m(\theta,\phi)$.

The addition theorems discussed so far are of the two-range form. In the series expansions one has to distinguish between $r_<$ and $r_>$ just as it is the case in the well-known Laplace expansion of the Coulomb potential in terms of spherical harmonics,

$$(|\vec{r}-\vec{R}|)^{-1} = \sum_{\ell,m} \left[4\pi/(2\ell+1)\right] r_<^\ell \, r_>^{-\ell-1} \, Y_\ell^{m*}(\Omega_{\vec{r}}) \, Y_\ell^m(\Omega_{\vec{R}}), \tag{2.5a}$$

$$r_< = \min(r,R), \quad r_> = \max(r,R). \tag{2.5b}$$

The two-range addition theorems make it possible to execute the integration over the angular variables. Unfortunately, the remaining radial integrals are more difficult. The fact that the expansions under consideration assume different forms in different regions of space makes the performance of the integrations sometimes very complicated as indefinite integrals of special functions will be needed in the general case. However, these indefinite integrals are usually very complicated and may even be unknown. In addition, not always a division of space can be chosen to fit the regions of convergence to the integration limits.[12]

A generalization of Eq. (2.1) is the bipolar expansion

$$f(\vec{r}_1+\vec{r}_2+\vec{r}_3) = \sum_{\ell_1\ell_2\ell_3} \sum_{m_1m_2m_3} f_{\ell_1\ell_2\ell_3}^{m_1m_2m_3}(r_1,r_2,r_3) \, Y_{\ell_1}^{m_1}(\Omega_{\vec{r}_1}) \, Y_{\ell_2}^{m_2}(\Omega_{\vec{r}_2}) \, Y_{\ell_3}^{m_3}(\Omega_{\vec{r}_3}) \tag{2.6}$$

which for $\vec{r}_1 = \vec{r}_A$, $\vec{r}_2 = -\vec{r}_B$, $\vec{r}_3 = -\vec{R}_{AB}$ distinguishes two centers A and B. For a division of space according to

$$|\vec{r}_i| < |\vec{r}_{i+1}+\ldots+\vec{r}_\mu|, \qquad i = 1,2,\ldots,\mu-1, \tag{2.7}$$

we can give bi- and polypolar expansions for certain functions

$$f(\vec{r}_1+\vec{r}_2+\ldots+\vec{r}_\mu), \tag{2.8}$$

especially for functions $r^n \, Y_\ell^m(\theta,\phi)$[6], which contain the Coulomb operator as a special case. Due to the division of space, however, these expansions are more suitable for investigations of molecular interactions than for integral evaluations.[13]

3. ETO CONVOLUTION AND OTHER TWO-CENTER INTEGRALS

The B-functions seem to fulfill several useful relationships just as the well-studied Bessel functions.

The addition theorems discussed in the foregoing enabled us to derive a convolution theorem for ETO's, which exhibits a remarkably simple structure. For B-functions with equal scaling parameters it is given by

$$\int d\vec{r}\; B_{N_1,L_1}^{M_1^*}(\alpha\vec{r})\; B_{N_2,L_2}^{M_2}[\alpha(\vec{r}-\vec{R})] =$$

$$= 4\pi\alpha^{-3}(-1)^{L_2} \sum_{\ell} \langle L_2M_2|L_1M_1|\ell m\rangle \sum_{t} (-1)^t \binom{\Delta\ell}{t} B^{m}_{N_1+N_2+L_1+L_2-\ell-t+1,\ell}(\alpha\vec{R}), \tag{3.1a}$$

$$\Delta\ell = (L_1+L_2-\ell)/2 \tag{3.1b}$$

A similar formula results for unequal scaling parameters.[14,15]

By linear transformations, we can transform the ETO's into each other, namely STO's, B-functions, and the Λ-functions defined in Section 4.[16] Therefore, relationships for B-functions are of direct use for the other ETO's as well, and for the time being it is sufficient to consider B-functions.

It should be noted that for variational calculations B-functions can be used instead of STO's because of the aforementioned transformation properties.[16] We developed programs for the overlap integrals of B-functions[17]. These programs were applied by U. Röbler and H. Grötsch to do band structure calculations of Si and Ge based upon the extended Hückel theory.[18,19]

The convolution theorem of B-functions enabled us to derive extremely compact formulas for the *two-center one-electron (i.e. overlap and nuclear attraction) integrals*[14,15] *as well as the Coulomb*[15] *and two-center hybrid integrals.*[8]

The *nuclear* attraction *integrals* with a one-center charge distribution can directly be reduced to a series of B-functions, whereas the *nuclear attraction integral* with a two-center charge distribution can be transformed into a linear combination of convolution integrals over B-functions. Now the *Coulomb integral* can be written as the convolution of a nuclear attraction integral and a B-function. Using our representation of the nuclear attraction integral as a sum of B-functions, the Coulomb integral can thus be reduced to convolution integrals over B-functions.[15]

Because of

$$\left(\frac{\partial}{\partial \vec{r}}\right)^2 B_{N,L}^M(\alpha\vec{r}) = \alpha^2 \left[B_{N,L}^M(\alpha\vec{r}) - B_{N-1,L}^M(\alpha\vec{r})\right] \tag{3.2}$$

the matrix elements of the *kinetic energy* operator can be reduced to convolution integrals.[14]

It should be noted that the formulas derived in this context are applicable for very high quantum numbers.

4. ORTHOGONAL AND NON-ORTHOGONAL COMPLETE SYSTEMS

Despite the intriguing properties of the B-functions, especially their convolution properties, we have found it essential to base our investigations not only on these and similar ETF sets, which form *non-orthogonal complete systems (NOCS)* but to use for our expansions also sets of ETF's which represent complete orthonormal systems (CONS), i.e. the Λ-functions, or equivalent biorthogonal systems.[16,20] The reason is that CONS have often valuable advantages over non-orthogonal systems especially if used in expansions, as, for instance, addition theorems. Addition theorems are expansions of a given function in terms of other functions defined with respect to another origin. If these functions form a *NOCS* (infinitely) many forms of such addition theorems are possible, and in fact, are continuously being published. On the other hand, CONS lead to unique expansions which have further advantages.

As only finite parts of infinite series can be used, we face some approximation problems. The Nth approximation $\tilde{f}_N$ of a given function $f \in L^2(\mathbb{R}^3)$ by a linear combination of functions $\phi_n \in L^2(\mathbb{R}^3)$, which form a set of basis functions $\{\phi_n\}_{n=1}^{\infty}$, is given by

$$\tilde{f}_N = \sum_{n=1}^{N} c_n^N \phi_n . \tag{4.1}$$

The coefficients c_n^N are determined by minimization of the norm

$$\|f - \tilde{f}_N\| \to \min \tag{4.2}$$

(least square fit).

The basis set $\{\phi_n\}_{n=1}^{\infty}$ is complete in $L^2(\mathbb{R}^3)$ if for any $\varepsilon > 0$ there exists an $N \in \mathbb{N}$ that

$$\|f - \tilde{f}_N\| < \varepsilon . \tag{4.3}$$

However, even for complete sets, the limit

$$\lim_{N\to\infty} c_n^N = c_n \tag{4.4}$$

need not exist, i.e. the existence of the formal expansion

$$f = \sum_{n=o}^{\infty} c_n \phi_n \tag{4.5}$$

is not guaranteed, and the coefficients c_n^N in Eq. (4.1) may diverge. Therefore, the properties of the function system $\{\phi_n\}_{n=1}^{\infty}$ are to be defined more strictly.

The basis set $\{\phi_n\}_{n=1}^{\infty}$ is exactly complete if omission of a single element ϕ_k makes the complete basis $\{\phi_n\}_{n-1}^{\infty}$ incomplete. The basis set $\{\phi_n\}_{n=1}^{\infty}$ is overcomplete if omission of a single element does not make the basis $\{\phi_n\}_{n=1,n\neq k}^{\infty}$ incomplete.[21] Exactly complete are CONS, like Λ-functions, whereas overcomplete are STO's, GTO's, RBF's, and all LCAO basis sets. Overcompleteness implies linear dependence. In this case, all the expansion coefficients c_n^N in Eq. (4.1) do depend on the order N of approximation, i.e. in general

$$c_n^N \neq c_n^{N+1}, \tag{4.6}$$

a very impractical condition. Also, the expansion coefficients need not have a finite limit for $N \to \infty$. This may lead to numerical catastrophes as can be seen in various examples;[22,23] see Table 1.

On the other hand, completeness and orthogonality implies exact completeness. If, therefore, $\{\psi_n\}_{n=1}^{\infty}$ is a CONS,

$$\langle\psi_n,\psi_m\rangle = \delta_{n,m}, \tag{4.7}$$

the expansion coefficients in Eq. (4.1) do not depend on the order N of the approximation. Improving the approximation, apart from a possible normalization change, the set of coefficients remains unchanged. This is a great numerical advantage. Hence,

$$f = \sum_{n=o}^{\infty} c_n \psi_n, \tag{4.8}$$

Table 1
Single-center calculations of H_2^+. The wavefunction is expanded in (a) terms of B-functions or (b) in terms of Λ-functions in the center of the molecule, respectively. The expansion coefficients in terms of B-functions diverge.[23]

L	N	Coefficients of B-functions $N_{max} = 6$	$N_{max} = 10$	$N_{max} = 16$	Coefficients of Λ-functions $N_{max} = 6$	$N_{max} = 10$	$N_{max} = 16$	L	N
0	1	$.318 \times 10^{-1}$	$.519 \times 10^{-2}$	$.174 \times 10^{-1}$	$.658 \times 10^{0}$	$.658 \times 10^{0}$	$.648 \times 10^{0}$	0	1
0	3	$.655 \times 10^{1}$	$.145 \times 10^{2}$	$.110 \times 10^{3}$	$.341 \times 10^{0}$	$.341 \times 10^{0}$	$.341 \times 10^{0}$	0	3
0	6	$-.829 \times 10^{1}$	$-.130 \times 10^{4}$	$-.175 \times 10^{6}$	$-.196 \times 10^{-1}$	$-.228 \times 10^{-1}$	$-.228 \times 10^{-1}$	0	6
0	9	-	$.735 \times 10^{3}$	$.682 \times 10^{7}$	-	$.356 \times 10^{-4}$	$.589 \times 10^{-3}$	0	9
0	12	-	-	$-.153 \times 10^{8}$	-	-	$.640 \times 10^{-3}$	0	12
0	15	-	-	$.127 \times 10^{7}$	-	-	$-.736 \times 10^{-4}$	0	15
0	16	-	-	$-.156 \times 10^{6}$	-	-	$-.651 \times 10^{-3}$	0	16
2	1	$.640 \times 10^{-1}$	$.321 \times 10^{0}$	$.378 \times 10^{0}$	$.179 \times 10^{0}$	$.179 \times 10^{0}$	$.179 \times 10^{0}$	2	3
2	3	$-.417 \times 10^{1}$	$.230 \times 10^{2}$	$.505 \times 10^{2}$	$-.249 \times 10^{-2}$	$-.227 \times 10^{-2}$	$-.228 \times 10^{-2}$	2	5
2	6	$.252 \times 10^{1}$	$.615 \times 10^{3}$	$.205 \times 10^{6}$	$.563 \times 10^{-2}$	$-.476 \times 10^{-4}$	$-.390 \times 10^{-4}$	2	8
2	9	-	-	$-.779 \times 10^{7}$	-	$.250 \times 10^{-2}$	$.186 \times 10^{-2}$	2	11
2	12	-	-	$.160 \times 10^{8}$	-	-	$-.755 \times 10^{-3}$	2	14
2	15	-	-	$-.123 \times 10^{7}$	-	-	$-.829 \times 10^{-3}$	2	17
2	16	-	-	$.148 \times 10^{6}$	-	-	$.976 \times 10^{-3}$	2	18

and the coefficients

$$c_n = \langle\psi_n | f\rangle, \tag{4.9}$$

which are scalar products, are bounded due to Parseval's identity[24]

$$\sum_{n=1}^{\infty} |c_n|^2 = \| f \| = (\langle f | f\rangle)^{1/2}. \tag{4.10}$$

Because of these considerations, we also looked into ETF's which form a CONS. With the help of the generalized Laguerre polynomials[10]

$$L_n^{\alpha}(x) = \sum_{m=o}^{\infty} (-1)^m \binom{n+\alpha}{n-m} \frac{x^m}{m!}, \tag{4.11}$$

we define the following set of functions:

$$\Lambda_{n,\ell}^{m}(\vec{r}) = \mathcal{N}_{n,\ell}\, e^{-r}\, L_{n-\ell-1}^{2\ell+2}(2r)\, \mathcal{Y}_{\ell}^{m}(2\vec{r}), \tag{4.12a}$$

where the normalization factor $\mathcal{N}_{n,\ell}$ is given by

$$\mathcal{N}_{n,\ell} = 2^{3/2}\, [(n-\ell-1)!/(n+\ell+1)!]^{1/2}. \tag{4.12b}$$

This set forms a complete orthonormal system in the Hilbert space $L^2(\mathbb{R}^3)$, i.e.:

$$\int d\vec{r}\, \Lambda_{n_1,\ell_1}^{m_1 *}(\vec{r})\, \Lambda_{n_2,\ell_2}^{m_2}(\vec{r}) = \delta_{n_1,n_2}\, \delta_{\ell_1,\ell_2}\, \delta_{m_1,m_2}. \tag{4.13}$$

Accordingly we define the following sets of functions:

$$\Phi_{-1,\ell}^{m}(r) = (2r)^{-1}\, e^{-r} \mathcal{Y}_{\ell}^{m}(2\vec{r}), \tag{4.14a}$$

$$\Phi_{n,\ell}^{m}(r) = e^{-r}\, L_n^{2\ell+1}(2r) \mathcal{Y}_{\ell}^{m}(2\vec{r}), \tag{4.14b}$$

$$\tilde{\Phi}_{n,\ell}^{m}(r) = [n!/(n+2\ell+1)!]\, (2r)^{-1}\, e^{-r}\, L_n^{2\ell+1}(2r) \mathcal{Y}_{\ell}^{m}(2\vec{r}). \tag{4.15}$$

These two sets are biorthogonal, i.e.:

$$(\tilde{\Phi}^{m}_{n,\ell}, \Phi^{m'}_{-1,\ell'}) = (2\beta)^{-3} \, [(2\ell)!n!/(2\ell+n+1)!] \, \delta_{\ell,\ell'} \, \delta_{m,m'}, \qquad (4.16a)$$

$$(\tilde{\Phi}^{m}_{n,\ell}, \Phi^{m'}_{n',\ell'}) = (2\beta)^{-3} \, \delta_{n,n'} \, \delta_{\ell,\ell'} \delta_{m,m'}. \qquad (4.16b)$$

We are able to transform all considered ETF's, like STO's, RBF's, Λ-functions into each other. Therefore, we can make use of the convolution properties of the B-functions and the orthogonality properties of the Λ-functions. This led us to one-range addition theorems for ETF's like B- and Λ-functions which do not require to distinguish between any $r_<$ and $r_>$ and, therefore, do not exhibit different forms according to the region in space just considered. For Λ-functions we have [16]

$$\Lambda^{m_1}_{n_1\ell_1}(\alpha(\vec{r}-\vec{R})) = {}^{-3/2} \sum_{\ell_3} \sum_{\ell_2} \sum_{m_2} \langle \ell_1 m_1 | \ell_3 m_1 - m_2 | \ell_2 m_2 \rangle \cdot (-1)^{\ell_2} \cdot$$

$$\cdot \sum_{n_2} \sum_{n_3} T^{n_1\ell_1, n_2\ell_2}_{n_3\ell_3} \Lambda^{m_1-m_2}_{n_3\ \ell_3}(\alpha\vec{r}) \, \Lambda^{m_2}_{n_2\ell_2}(\alpha\vec{R}); \qquad (4.17a)$$

$$\ell_2 + 1 \leq n_2 < \infty, \qquad (4.17b)$$

$$\max(\ell_3 + 1, |n_1-n_2| - 1) \leq n_3 \leq n_1 + n_2 + 1. \qquad (4.17c)$$

This series expansion is convergent not only with respect to the norm of $L^2(\mathbb{R}^3)$ but also in a pointwise sense.[16]

5. ONE-RANGE ADDITION THEOREMS AND SINGLE-CENTER EXPANSIONS

After having introduced the Λ-functions in our derivations, we found out about independent work of other groups. Already in 1966 Smeyers[25] investigated the evaluation of 2- and 3-center integrals over STO's by shifting one orbital from its center to another center and by using an expansion of the shifted function in terms of Λ-functions. The 2-center integrals remaining in the formula for the reduction of a 3-center integral, were then evaluated in elliptical coordinates. In 1978 and 1980 Guseinov[26,27] reformulated these suggestions but gave analytical results in a more complicated form. In this method the expansion coefficients, which are themselves overlap integrals between Λ-functions and STO's, are evaluated as linear combinations of overlap integrals over STO's. However, such a representation of a set of orthonormal (Λ-)functions as a linear combination of STO's will lead to serious numerical difficulties due to the reasons discussed above, especially if higher quantum numbers are taken into account. We, therefore, think it is preferable to stay within the Λ-set if this is once introduced, and we are able to do so because we got the necessary formulas.

Our addition theorem for Λ-functions, Eq. (4.17), has several advantages suitable for molecular integral evaluations: (a) The expansion coefficients T are explicitely given for any order of the term, they are independent on the geometry of the molecule, they are calculated and programmed once and for all[28] ; (b) the expansion requires no division of space but the same expansion is valid over the whole space. This is very advantageous for the integration procedure, because the limit of the integrals remains 0 or ∞, no division of space is necessary and, therefore, no indefinite integrals are to be evaluated; (c) the expansion holds and is practical for very high quantum numbers.

The addition theorems are not only useful for integral evaluations, but also in other contexts requiring the shifting of functions. One side aspect is the possibility to improve the *single-center expansion (SCE) method.*[23]

The regular SCE-method that is straightforward, has the following features: (a) No information about the position of the nuclei is incorporated in the ansatz, rather the molecular structure is to be elaborated by the Rayleigh-Ritz variational (RRV-) method; (b) the molecular wave function is expanded in terms of a complete set of basis functions (may be a CONS) about one arbitrary point; (c) due to (b), no multicenter integrals do occur; (d) one observes slow convergence; (e) a big eigenvalue problem results; (f) high quantum numbers cause additional problems as instabilities, deterioration of convergence, etc. . The points (b,c) are in favor of the SCE-method, the points (a,d,e,f) discredit this procedure.

On the other hand, in the LCAO-MO-method (a) information about the position of the nuclei (i.e. the molecular structure) is already incorporated in the ansatz; (b) the function set is overcomplete, the functions become linearly dependent if higher AO's are included; (c) complicated multicenter integrals are produced; (d) good convergence of the procedure can be obtained; (e) a moderate eigenvalue problem results; (f) problems with high quantum numbers do occur, partly because of (b). The points (a,d,e) name advantages of the method, whereas the points (b,c,f) specify its serious drawbacks.

Now, the Λ-addition theorem Eq. (4.17) can be used to incorporate the information about the molecular structure also into the ansatz for the SCE-method. The AO's are expressed in terms of Λ-functions, and these are shifted to the single center 0. As one can use only finite expansions, each AO is represented by a *contracted* set of Λ-functions centered at 0. Using additional "free" Λ-functions at 0 as usual in the SCE-method, one has a refined procedure that leads to a much smaller eigenvalue problem and reaches results faster. It also makes it possible to go to rather high quantum numbers. Therefore, the disadvantages of the points listed for SCE's may be suppressed by combining this method with some merits of the LCAO-method. SCE-calculations with Λ-functions on H_2^+ and comparisons with the exact solution justify these statements.[23] If executed with a *NOCS*. These calculations also illustrate nicely the general statements given above and in Section 4.

It should be noted that in quantum chemistry we usually determine an *unknown* function ψ via the Rayleigh-Ritz variational (RRV-) method by minimizing

$$E[\psi] = \frac{\langle\psi|H|\psi\rangle}{\langle\psi|\psi\rangle} \,. \tag{5.1}$$

For the matrix elements to exist, ψ must be an element of the Sobolev space $W_2^{(1)}$.[29,30] If we expand ψ in terms of a complete set $\{\phi_n\}_{n=1}^{\infty}$, completeness in L^2 is not sufficient, but completeness in $W_2^{(1)}$ is required. Otherwise, the variational procedure may converge to a wrong limit.[31] In order to reduce the linear dependence of the basis under consideration, often some functions are deleted. However, after omission of some functions, the basis may no longer be complete in $W_2^{(1)}$. Then, convergence to a wrong limit may occur.[31] Hence, omission of some functions must be handled with care.

6. THREE- AND FOUR-CENTER INTEGRALS

6.1. Expansions of Orbitals and Operators

In order to avoid the difficulties due to the otherwise necessary division of integration space, it seems recommendable to use one-range addition theorems and orthogonal expansions for the evaluation of three- and four-center ETO integrals.[32]

The *three-center exchange integral*

$$I_{ABAC} = \int d\tau_1 \int d\tau_2 \; \chi_A(1) \; \chi_B(1) \; \frac{1}{r_{12}} \; \chi_A(2) \; \chi_C(2) \tag{6.1}$$

over STO's (or B-functions) can be transformed into a series of the type

$$I_{ABAC} = \sum_{\mu\upsilon} c_{\mu\upsilon} \; S_\mu(\vec{R}_{AB}) \; S_\upsilon(\vec{R}_{AC}) \tag{6.2}$$

with analytically given expansions coefficients $c_{\mu\upsilon}$ and convolution integrals

$$S_\mu(\vec{R}_{AB}) = \langle\Lambda_A|\chi_B\rangle, \qquad S_\upsilon(\vec{R}_{AC}) = \langle\Lambda_A|\chi_C\rangle, \tag{6.3}$$

if one expands the "operator"

$$\frac{e^{-r_1}\,e^{-r_2}}{|\vec{r}_1-\vec{r}_2|} = \sum_{n\ell m}\ \sum_{n'\ell'm'} \text{coeff.} \times \Lambda^{m}_{n\ell}(\vec{r}_1)\ \Lambda^{m'}_{n'\ell'}(\vec{r}_2) \tag{6.4}$$

in terms of Λ-functions.

A calculation with 1s-orbitals[16] for the two-center case ($\vec{R}_{AB} = \vec{R}_{AC} = 1.0\hat{z}$) shows good convergence and coincidence with Hirschfelder's[33] result 0.436651. In Table 2a we plot the results we obtain for the maximal summation indices, i.e. "quantum numbers" of Λ-functions included. In Table 2b we plot the equivalent results we obtain with 1s-orbitals for the three-center collinear case ($\vec{R}_{AC} = -\vec{R}_{AB} = 1.0\hat{z}$). Here, $\hat{z}$ represents the unit vector in z-direction.

Table 2a
2-center exchange integrals with 1s-orbitals; $\vec{R}_1 = \vec{R}_2 = 1.0\hat{z}$
Hirschfelder's result : 0,436651

ℓ^{max}	$n^{max} = 10$	$n^{max} = 20$	$n^{max} = 30$
4	0.4366179	0.4366460	0.4366466
6	0.4366181	0.4366488	0.4366508
8	0.4366181	0.4366488	0.4366508

Table 2b
3-center exchange integral with 1s-orbitals; $\vec{R}_1 = -\vec{R}_2 = 1.0\ \hat{z}$.

ℓ^{max}	$n^{max} = 10$	$n^{max} = 20$	$n^{max} = 30$
4	0.3953624	0.3953586	0.3953585
6	0.3953622	0.3953563	0.3953557
8	0.3953622	0.3953563	0.3953557

A similar approach is possible for the *four-center integral*

Table 3a
2-center nuclear attraction (Op = 1/r) and overlap (Op = 1) integrals $\langle \Lambda_{n_b \ell_b}^{m_b}(\alpha\vec{r})|Op|\Lambda_{n_k \ell_k}^{m_k}(\vec{r}-\vec{R}_k)\rangle$ for $\vec{R}_k = 2.0\ \hat{z}$ (at α = 5.0, 10.0, 15.0) as a function of α
Trivedi and Steinborn[28]

$n_b \ell_b$	$n_k \ell_k$	Op	$\vec{R}_k = 2.0\ \hat{z}$		
			α = 5.0	α = 10.0	α = 15.0
1s	1s	1	9.6329658064317 D-2	3.422686919 D-2	1.8635375 D-2
			23	30	23
		r^{-1}	2.4167833534611 D-1	1.7116963083 D-1	1.39770890 D-1
			23	42	42
2s		1	-1.6445257772272 D-1	-5.92332905 D-2	-3.2272210 D-2
			28	31	41
		r^{-1}	-1.3756038423565 D-1	-9.87433455 D-2	-8.068387 D-2
			25	43	59
2p		1	7.1294652786865 D-2	1.3472620 D-2	4.935295 D-3
			43	38	42
		r^{-1}	9.1992908868526 D-2	2.2874016 D-2	1.855146 D-2
			30	49	65
1s	2s	1	-4.7995945073170 D-2	-1.899539641 D-2	-1.0570467 D-2
			25	51	53
		r^{-1}	-1.2940921971276 D-1	-9.688948736 D-2	-7.9985959 D-2
			31	62	52
2s		1	6.763391625685 D-2	3.1166275 D-2	1.7876613 D-2
			33	41	42

Table 3a Continued

$n_b \ell_b$	$n_k \ell_k$	Op	$\vec{R}_k = 2.0\ \hat{z}$		
			$\alpha = 5.0$	$\alpha = 10.0$	$\alpha = 15.0$
2p		r^{-1}	5.2961374569700 D-2 34	5.1522430 D-2 54	4.454612 D-2 75
		1	4.7864746162986 D-2 31	8.212428 D-3 52	2.92314 D-3 42
		r^{-1}	5.9561632228508 D-2 33	2.02979771 D-2 52	1.089684 D-2 75

$$I_{ABCD} = \int d\tau_1 \int d\tau_2 \; \chi_A(1) \;\; \chi_B(1) \frac{1}{r_{12}} \; \chi_C(2) \;\; \chi_D(2), \tag{6.5}$$

which leads to a series of products of Coulomb integrals $C^{\ell\ell'}_{nn'}(\vec{R}_{AD})$ and overlap integrals $S_{n\ell}(\vec{R}_{AB})$ etc. according to

$$I_{ABCD} = \sum_{n\ell} \sum_{n'\ell'} C^{\ell\ell'}_{nn'}(\vec{R}_{AD}) \cdot S_{n\ell}(\vec{R}_{AB}) \cdot S_{n'\ell'}(\vec{R}_{DC}), \tag{6.6}$$

if one uses the expansion

$$\frac{e^{-r_1} \; e^{-r_2}}{|\vec{R}_{AD}+\vec{r}_2-\vec{r}_1|} = \sum_{n\ell} \sum_{n'\ell'} \sum_{mm'} C^{\ell\ell'}_{nn'}(\vec{R}_{AD}) \cdot \Lambda^{m}_{n\ell}(\vec{r}_1) \;\; \Lambda^{m'}_{n'\ell'}(\vec{r}_2) \tag{6.7}$$

In order to compare alternative strategies, we discuss finally the most general one-electron integral, i.e. the *three-center nuclear attraction (NA) integral*

$$D = \int d\vec{r} \;\; \chi_A(\vec{r}-\vec{R}_A) \frac{1}{r} \; \chi_B(\vec{r}-\vec{R}_B). \tag{6.8}$$

Although compact expressions can be derived[12], we want to investigate the application of the Λ-addition theorem. As the first step towards evaluation, one can shift one of the two basis functions - say, the bra χ_A - to the center of attraction in order to obtain a series of two-center NA integrals.[15] These integrals may further be reduced to a series of one-center NA integrals by shifting the ket from $\vec{R}_B$ to the common center of nuclear attraction and the bra. For χ_A and χ_B representing Λ-functions with scaling parameters α and β, respectively, the results for these two-center NA integrals are given in Tables 3a,b[28]. Some results for the three-center NA integrals are given in Tables 4,5[28]. For an accuracy of 10^{-6} one may need up to 100 terms in a sum. This, however, causes no problems and no numerical instabilities were encountered. The number of terms can be cut down by applying the Λ-addition theorem only once and using the Laplace expansion for nuclear attraction operator.

6.2. INTEGRALS EXPANDED INTO FUNCTIONS OF THE NUCLEAR DISTANCES

Multicenter integrals are square integrable functions of the inter-nuclear distances.[20] For the *three-center NA integral* we have $D = D(\vec{R}_A,\vec{R}_B)$. If this function is expanded in terms of the biorthogonal basis defined in Eq. (4.14)

Table 3b

2-center nuclear attraction ($Op = 1/r$) and overlap integrals ($Op = 1$) for $\alpha = 10.0$ as a function of $\vec{R}_k$ (at $\vec{R}_k = 2.0\ \hat{z}$, $5.0\ \hat{z}$, $10.0\ \hat{z}$)

Trivedi and Steinborn[28]

$n_b \ell_b$	$n_k \ell_k$	Op	$\alpha = 10.0$					
			$\vec{R}_k = 2.0\ \hat{z}$		$\vec{R}_k = 5.0\ \hat{z}$		$\vec{R}_k = 10.0\ \hat{z}$	
1s	1s	1	3.422686919 30	D-2	1.7251366 20	D-3	1.167123 19	D-5
		r^{-1}	1.7116963083 42	D-1	8.57420948 31	D-3	5.7889755 25	D-5
2s		1	-5.92332905 31	D-2	-3.035725 30	D-3	-2.06492 23	D-5
		r^{-1}	-9.87433455 43	D-2	-5.068194 37	D-3	-3.44997 43	D-5
2p		1	1.3472620 38	D-2	6.92482 43	D-4	4.70316 23	D-6
		r^{-1}	2.2874016 49	D-2	1.722326 37	D-3	1.16665 37	D-5
1s	2s	1	-1.899539641 51	D-2	-6.9081270 21	D-3	-1.140638 19	D-4
		r^{-1}	-9.688948736 62	D-2	-3.44932197 37	D-2	-5.669730 29	D-4
2s		1	3.1166275 41	D-2	1.2008190 31	D-2	2.00696 26	D-4
		r^{-1}	5.1522430 54	D-2	2.0024171 43	D-2	3.350695 37	D-4
2p		1	8.212428 52	D-3	-1.960636 35	D-3	-4.04358 30	D-5
		r^{-1}	2.02979771 52	D-2	-4.908194 43	D-3	-1.005463 51	D-4

Table 4

3-Center nuclear attraction integral $<\Lambda^{0}_{10}(\vec{r}-\vec{R}_1)\ |\frac{1}{r}|\ \Lambda^{0}_{10}(\vec{r}-\vec{R}_2)>$ for $\vec{R}_1 = 2.5\ \hat{z}$, $\vec{R}_2 = 0.5\ z$; as a function of the upper limits ℓ_2^{max} and n_2^{max} on the expansion indices ℓ_2 (angular momentum) and n_2 (principal quantum number) of Eq. (4.17). Trivedi and Steinborn.[28] [The converged value is 0.341022 (Hirschfelder and Weygandt, 1938)]

ℓ_2^{max} \ n_2^{max}	10	15	20	25	30
0	0.274725	0.275633	0.275324	0.275425	0.275461
1	0.333420	0.333906	0.333660	0.333751	0.333786
2	0.339719	0.340530	0.340181	0.340301	0.340332
3	0.340390	0.341154	0.340861	0.340941	0.340984
4	0.340481	0.341210	0.340922	0.341018	0.341046
5	0.340486	0.341222	0.340926	0.341023	0.341057

Table 5

3-center nuclear attraction integrals for two linear geometries. Both orbitals are 1s states with unit orbital exponent; the center of attraction is at the origin. Trivedi and Steinborn[28]

$\vec{R}_b$	$\vec{R}_k$	n_2^{max}	ℓ_2^{max}	Integral
$1.5\ \hat{z}$	$-0.5\ \hat{z}$	30	6	0.4499 (±1)
		150	20	0.449956(±1)
$2.5\ \hat{z}$	$0.5\ \hat{z}$	30	6	0.34106 (±4)
		150	20	0.341022(±1)

$$D(\vec{R}_A,\vec{R}_B) = \sum_{n\ell m} \langle\tilde{\Phi}_{n\ell}|D\rangle_{\vec{R}_A} \; \Phi^m_{n\ell}(\vec{R}_A), \qquad (6.9)$$

the expansion coefficients are scalar products

$$\langle\tilde{\Phi}^m_{n\ell}|D\rangle_{\vec{R}_A} = \int d\vec{R}_A \; \Phi^m_{n\ell}(\vec{R}_A)^* \; D(\vec{R}_A,\vec{R}_B) \qquad (6.10)$$

With the help of the convolution theorem, they can be represented by a finite sum of overlap integrals $S(\vec{R}_B)$ of B-functions.

For a basis set of functions defined by Eq. (4.14), we obtained by this procedure satisfying results with good convergence, as shown in Table 6 for the three-center NA integral with 1s-functions. However, it would be preferable to use Λ-functions, as the scalar products which occur in Eq. (6.10) can then be evaluated directly.[16]

Table 6
Partial sums of the series Eq. (6.9); $\vec{R}_1 = 1.5\,\hat{z}$, $\vec{R}_2 = -0.5\,\hat{z}$.
Hirschfelder's result 1938: 0.449956

ℓ^{max}	$n^{max} = 4$	$n^{max} = 8$	$n^{max} = 10$
0	0.525805	0.525763	0.525776
3	0.449979	0.449856	0.449857
5	0.450067	0.449952	0.449950

The same procedure can be applied to *four-center integrals*. In this case, the expansion coefficients are finite sums of Coulomb integrals for which compact results are known.[15]

Hence, various procedures are available to finally evaluate the ETO molecular multicenter integrals.

ACKNOWLEDGMENTS

It is a pleasure to acknowledge cooperation and discussions with Dr. E. Filter, Dr. H. H. Kranz, Dr. H. Trivedi, and Mr. E. J. Weniger, who contributed essential results as cited. Meetings with Dr. David M. Silver made possible by NATO Research Grant No. 1861 were very helpful and stimulating. The work reported here was supported in part by Deutsche Forschungsgemeinschaft, Fonds der chemischen Industrie, and NATO Scientific Affairs Division. The computations were done at the Rechenzentrum der Universität Regensburg. Thanks are due to its staff for excellent support.

The author is very much indebted to Professor Herbert W. Jones, Director, and Dr. C.A. Weatherford, for inviting him to present the results, which are summarized in this paper, at the International Conference on ETO Multicenter Molecular Integrals, at Tallahassee, Florida, in August 1981.

REFERENCES

1. Boys, S.F.: 1950, Proc. Roy. Soc. A 200, p.542.

2. Steinborn, E.O. and Filter, E: 1979, Theoret. Chim. Acta (Berl.) 52, p.189.

3. Fieck, G.: 1980, Habil. Arbeit, Fak. Chemie, Universität Regensburg.

4. Steinborn, E.O. and Ruedenberg, K.: 1973, Adv. Quantum Chem. 7, p.1.

5. Steinborn, E.O. and Filter, E.: 1975, Theoret. Chim. Acta (Berl.) 38, p.247.

6. Steinborn, E.O. and Filter, E.: 1975, Int. J. Quantum Chem. Symp. 9, p.435.

7. Steinborn, E.O. and Filter, E.: 1975, Theoret. Chim. Acta (Berl.) 38, p.273.

8. Filter, E.: 1978, PhD-Thesis, Fak. Chemie, Universität Regensburg.

9. Shavitt, I., In: Alder, B., Fernbach, S. and Rotenberg, M. (Eds.): 1973, Methods in Computational Physics, Academic Press, New York.

10. Magnus, W., Oberhettinger, F. and Soni, R.P.: 1966, Formulas and Theorems for the Special Functions of Mathematical Physics, Springer-Verlag, New York.

11. Condon, E.U. and Shortley, G.H.: 1935, Theory of Atomic Spectra, Cambridge U.P., Cambridge.

12. Steinborn, E.O. and Ruedenberg, K.: 1972, Intro. J. Quantum Chem. 6, p.413.

13. Hirschfelder, J.O., Curtiss, C.F. and Bird, R.B.: 1954, Molecular Theory of Gases and Liquids, Wiley, New York.

14. Filter, E. and Steinborn, E.O.: 1978, J. Math. Phys. 19, p.79.

15. Filter, E. and Steinborn, E.O.: 1973, Phys. Rev. A 18, p.1.

16. Filter, E. and Steinborn, E.O.: 1980, J. Math. Phys. 21, p.2725.

17. Weniger, E.J. and Steinborn, E.O.: to be published.

18. Röbler, E. and Grötsch, H.: to be published.

19. Grötsch, H.: 1981, Diplomarbeit, Fak. Physik, Universität Regensburg.

20. Steinborn, E.O. and Filter, E.: 1980, Int. J. Quantum Chem. 18, p.219.

21. Klahn, B. and Bingel, W.A.: 1977, Int. J. Quantum Chem. 11, p.943.

22. Löwdin, P.O.: 1970, Adv. Quantum Chem. 5, p.185.

23. Kranz, H.H. and Steinborn, E.O.: to be published.

24. Higgins, J.R.: 1977, Completeness and Basis Properties of Sets of Special Functions, Cambridge U.P., Cambridge.

25. Smeyers, Y.G.: 1966, Theoret. Chim. Acta (Berl.) 4, p.452.

26. Guseinov, I.I.: 1978, J. Chem. Phys. 69, p.4990.

27. Guseinov, I.I.: 1980, Phys. Rev. A 22, p.369.

28. Trivedi, H.P. and Steinborn, E.O.: submitted for publication.

29. Richtmeyer, R.D.: 1978, Principles of Advanced Mathematical Physics I, Springer-Verlag, New York.

30. Sobolew, S.L.: 1963, Applications of Functional Analysis in Mathematical Physics, American Mathematical Society, Providence, Rhode Island.

31. Klahn, B. and Bingel, W.A.: 1977, Theoret. Chim. Acta (Berl.) 44, p.27.

32. Filter, E. and Steinborn, E.O.: to be published.

33. Hirschfelder, J.O.: 1938, J. Chem. Phys. 6, p.806.

SCALED HYDROGENIC STURMIANS AS ETOs

C.A. Weatherford
Institute for Molecular Computations and Physics Department
Florida A&M University, Tallahassee, Florida 32307

Recently there have been renewed attempts to obtain analytical series expansions of multi-center molecular integrals over Exponential Type Orbitals (ETOs)[1-8]. In this regard, ETOs that exhibit radial as well as angular orthonormality have been exploited, first by Guseinov[5], then by Weatherford and Jain[6], and finally by Filter and Steinborn[8]. All three of these works have used the same basis sets - namely the orthonormal functions first discussed by Slater[9]. These functions may then properly be called orthonormal Slaters (ONSs). The radial parts may be constructed, as Slater points out, by an analytical Gram-Schmidt process from the minimum n-quantum number, normalized radial Slaters.

An important characteristic of the ONSs is that the screening constant is arbitrary. As a result of this and the orthonormality, all ETO multicenter integrals can be reduced to sums over equal screening constant, two-center overlaps. Guseinov then does the overlaps in elliptic coordinates. Filter and Steinborn[2], however have given a terminating, spherical coordinate, computable form for these overlaps. This formula was then used by Weatherford and Jain[6] to evaluate multi-center, electron-molecule interaction integrals. Filter and Steinborn[8] then exploited their own formula to accomplish the projection of the translation operator onto ONSs, thus also reducing the multi-center ETO integrals to equal screening constant overlaps, and thereby giving a compact form for the matrix elements of the translation operator.

In addition, because of the properties of the ONSs, Filter and Steinborn[8] have shown that normalized Slaters (NSs) and positive power reduced Bessel functions, (RBFs), can both be expanded in terminating series of ONSs. And similarly, ONSs can be expanded in terminating series of NSs or RBFs. Thus, the ONSs may be used to perform multi-center integrals over NSs or RBFs. It might be said that this set of ETOs are closed under the operation of taking finite linear combinations

C. A. Weatherford and H. W. Jones (eds.), ETO Multicenter Molecular Integrals, 29–34.

To the three ETOs mentioned above, the intent of the present work is to add a fourth ETO - namely, Hydrogenic Sturmians[10,11]. An expression for the expansion of ONSs in Hydrogenic Sturmians (HSs), and also for the inverse transformation may be given as follows:

$$\Phi_{n_i \ell_i m_i}(\alpha,\vec{r}) = \sum_{n_f=\ell_i+1}^{n_i} a_{n_f}^{n_i \ell_i}(\alpha)\, \mathcal{S}_{n_f \ell_i m_i}(\alpha,\vec{r}) \tag{1}$$

where

$$a_{n_f}^{n_i \ell_i}(\alpha) = \left[\frac{2\alpha n_f(\ell_i+1)(n_f-\ell_i-1)!\,(n_f+\ell_i)!}{(2n_i)!}\right]^{\frac{1}{2}} \times \sum_{\sigma=0}^{n_f-\ell_i-1} \frac{(-1)^{\sigma}(n_i+\ell_i+\sigma)!}{(n_f-\ell_i-1-\sigma)!\,(2\ell_i+1+\sigma)!\,\sigma!} \tag{2}$$

also

$$\mathcal{S}_{n_i \ell_i m_i}\left(\frac{\gamma}{\ell_i+1},\vec{r}\right) = \sum_{n_f=\ell_i+1}^{n_i} b_{n_f}^{n_i \ell_i}(\gamma)\, \Phi_{n_f \ell_i m_i}\left(\frac{\gamma}{\ell_i+1},\vec{r}\right) \tag{3}$$

where

$$b_{n_f}^{n_i \ell_i}(\gamma) = \left[\frac{(n_i-\ell_i-1)!\,(n_i+\ell_i)!}{\gamma n_i}\right]^{\frac{1}{2}} (-1)^{n_f-\ell_i-1} \frac{\left[(n_f)!\right]^{\frac{1}{2}}}{(n_i-n_f)!\,(n_f+\ell_i)!\,(n_f-\ell_i-1)!} \tag{4}$$

For completeness, I record the definitions of Φ and $\mathcal{S}$ along with their normality and orthogonality conditions

$$\Phi_{n\ell m}(\alpha,\vec{r}) \equiv \phi_n(\alpha,r)\, Y_\ell^m(\hat{r}) \tag{5}$$

$$\phi_n(\alpha,r) = N_n(\alpha)\, r^{n-1} e^{-\alpha r} \; ; \qquad N_n(\alpha) = \frac{(2\alpha)^{n+\frac{1}{2}}}{[(2n)!]^{\frac{1}{2}}} \tag{6}$$

$$\delta_{\ell\ell'}\ \delta_{mm'} = \int d\vec{r}\ \Phi^*_{n\ell'm'}(\alpha,\vec{r})\ \Phi_{n\ell m}(\alpha,\vec{r}) \tag{7}$$

$$\mathcal{S}_{n\ell m}\left(\frac{\gamma}{\ell+1},\vec{r}\right) \equiv S_{n\ell}\left(\frac{\gamma}{\ell+1},r\right) Y^m_\ell(\hat{r}) \tag{8}$$

$$S_{n\ell}\left(\frac{\gamma}{\ell+1},r\right) = \sum_{\sigma=0}^{n-\ell-1} s^\sigma_{n\ell}(\gamma)\ r^{\ell+\sigma}\ e^{-\left(\frac{\gamma}{\ell+1}\right)r} \tag{9}$$

$$s^\sigma_{n\ell}(\gamma) = \left[\frac{4\gamma^2(n-\ell-1)!}{n(\ell+1)^3(n+\ell)!}\right]^{\frac{1}{2}} (-1)^\sigma \frac{(n+\ell)!}{(n-\ell-1-\sigma)!\ (2\ell+1+\sigma)!\ \sigma\ !} \left[\frac{2\gamma}{\ell+1}\right]^{\sigma+\ell} \tag{10}$$

$$\delta_{nn'}\ \delta_{\ell\ell'}\ \delta_{mm'} = \int d\vec{r}\ \mathcal{S}^*_{n'\ell'm'}\left(\frac{\gamma}{\ell'+1},\vec{r}\right) \frac{1}{r} \mathcal{S}_{n\ell m}\left(\frac{\gamma}{\ell+1},\vec{r}\right) \tag{11}$$

Because of a variety of definitions of Laguerre polynomials, there may be some confusion about the equivalence of the functions used by Guseinov [1] and by Filter and Steinborn [8]. Both functions are the same, and I call them ONSs. To avoid confusion, I now give their relationship to NSs, along with their definition and orthonormality condition ($\mathcal{G}$ ≡ ONSs).

$$\Phi_{n_i\ell_i m_i}(\alpha,\vec{r}) = \sum_{n_f=\ell_i+1}^{n_i} C^{n_i\ell_i}_{n_f}\ \mathcal{G}_{n_f\ell_i m_i}(\alpha,\vec{r}) \tag{12}$$

where

$$C^{n_i\ell_i}_{n_f} = \left[\frac{(n_f-\ell_i-1)!\ (n_f+\ell_i+1)!}{(2n_i)!}\right]^{\frac{1}{2}} \times \sum_{\sigma=0}^{n_f-\ell_i-1} \frac{(-1)^\sigma (n_i+\ell_i+\sigma+1)!}{(n_f-\ell_i-1-\sigma)!\ (2\ell_i+2+\sigma)!\ \sigma\ !} \tag{13}$$

also

$$\mathcal{G}_{n_i\ell_i m_i}(\alpha,\vec{r}) = \sum_{n_f=\ell_i+1}^{n_i} d_{n_f}^{n_i\ell_i} \Phi_{n_f\ell_i m_i}(\alpha,\vec{r}) \tag{14}$$

where

$$d_{n_f}^{n_i\ell_i} = \left[(n_i-\ell_i-1)!\,(n_i+\ell_i+1)!\right]^{\frac{1}{2}} \times$$
$$\times \quad (-1)^{n_f-\ell_i-1} \frac{\left[(2n_f)!\right]^{\frac{1}{2}}}{(n_i-n_f)!\,(n_f+\ell_i+1)!\,(n_f-\ell_i-1)!} \tag{15}$$

Thus

$$\mathcal{G}_{n\ell m}(\alpha,\vec{r}) \equiv G_{n\ell}(\alpha,r)\, Y_\ell^m(\hat{r}) \tag{16}$$

$$G_{n\ell}(\alpha,r) = \sum_{\sigma=o}^{n-\ell-1} g_{n\ell}^{\sigma}(\alpha)\, r^{\ell+\sigma}\, e^{-\alpha r} \tag{17}$$

$$g_{n\ell}^{\sigma}(\alpha) \equiv \left[(2\alpha)^3 (n-\ell-1)!\,(n+\ell+1)!\right]^{\frac{1}{2}} (-1)^{\sigma} \frac{(2\alpha)^{\ell+\sigma}}{(n-\ell-1-\sigma)!\,(2\ell+2+\sigma)!\,\sigma\,!} \tag{18}$$

$$\delta_{nn'}\,\delta_{\ell\ell'}\,\delta_{mm'} = \int d\vec{r}\; \mathcal{G}^*_{n'\ell'm'}(\alpha,\vec{r})\; \mathcal{G}_{n\ell m}(\alpha,\vec{r}) \tag{19}$$

Comparing the orthogonality conditions of Φ, $\mathcal{S}$, $\mathcal{G}$ given by (7), (11), and (19) respectively, the Φ's are angularly orthogonal, but not radially. The $\mathcal{S}$'s have an inverse r weighting factor and the $\mathcal{G}$'s have a weighting factor of 1. Thus, the $\mathcal{S}$'s and $\mathcal{G}$'s are both angularly and radially orthogonal. It is then clear that the Hydrogenic Sturmians may be added to the set of ETOs that are closed under the operation of finite linear combinations.

It is also clear that translation formulae for HSs may be generated according to the following scheme

$$HS_a \rightarrow NS_a \rightarrow ONS_a \overset{\infty}{\rightarrow} ONS_b \rightarrow NS_b \rightarrow HS_b \tag{20}$$

where the actual translation is accomplished by the formula of Filter and Steinborn[8]. Note: 'a' and 'b' of (20) stand for different coordinate centers and $\rightarrow$ stands for a finite linear combination. Also $\overset{\infty}{\rightarrow}$ implies an infinite linear combination.

Finally, an important characteristic of the HSs is that the one-particle expectation values are equally spaced. In explanation

$$\bar{r}(i) \equiv \int d\vec{r}\, \mathcal{S}_i^*\left(\frac{\gamma}{\ell_i+1},\vec{r}\right)\, r\, \mathcal{S}_i\left(\frac{\gamma}{\ell_i+1},\vec{r}\right) \tag{21}$$

where $i=(n\ell m)_i$, defines the expectation value of r with respect to the ith HS. Thus

$$\bar{r}(i) = \int_0^\infty dr\, S_{n_i \ell_i}\left(\frac{\gamma}{\ell_i+1},r\right)\, r^3\, S_{n_i \ell_i}\left(\frac{\gamma}{\ell_i+1},r\right) \tag{22}$$

And then if

$$\Delta_{ij} \equiv \bar{r}(i) - \bar{r}(j) \tag{23}$$

its true that

$$\Delta_{ij} = \Delta_{mn} \quad \text{for all } i,j,m,n \tag{24}$$

This fact is useful because in order to span a sphere of configuration space, say from $r=0$ to $r=r_t$, with N HSs, only one value of γ needs to be calculated. That is, require

$$r_t = \sum_{i=1}^{N-1} \Delta_{i+1,i}(\gamma) \tag{25}$$

and solve for γ considering Δ as a function of γ. This is easily accomplished on a digital computer.

REFERENCES

1. Guseinov, I.I.: 1978, J. Chem. Phys. 69, p.4990.

2. Filter, E.F. and Steinborn, E.O.: 1978, Phys. Rev. A 18, p.1.

3. Jones, H.W. and Weatherford, C.A.: 1978, Int. J. Quantum Chem. Symp. 12, p.483.

4. Antolovic, D. and Delhalle, J.: 1980, Phys. Rev. A 21, p.1815.

5. Guseinov, I.I.: 1980, Phys. Rev. A 22, p.369.

6. Weatherford, C.A. and Jain, B.L.: 1980, Int. J. Quantum Chem. Symp. 14, p.493.

7. Jain, B.L. and Weatherford, C.A.: 1980, Bull. Am. Phys. Soc. 25, p.1142.

8. Filter, E. and Steinborn, E.O.: 1980, J. Math. Phys. 21, p.2725.

9. Slater, J.C.: 1960, Quantum Theory of Atomic Structure I, McGraw-Hill, New York.

10. Rotenberg, M.: 1962, Ann. Phys. 19, p.262.

11. Shakeshaft, R.: 1975, J. Phys. B.: Atom. Molec. Phys. 8, p.1114.

MULTI-CENTER INTEGRALS FROM THE GENERAL EXPANSION FORMULA DEVELOPED PREVIOUSLY FOR SLATER ORBITALS

R. R. Sharma
Department of Physics, University of Illinois, Chicago, Illinois 60680

Employing the general formula developed previously by the author for the expansion of a Slater type orbital from one center on to the other, the expressions for the two-center hybrid, Coulomb and exchange integrals and two-and three-center hyperfine interaction matrix elements have been presented. The derived hybrid and Coulomb integrals and two-center hyperfine interaction matrix elements have been obtained in closed forms. The importance of the formulas has been demonstrated by presenting calculations for Slater type orbitals involving s, p, d, f, g, h quantum numbers. The evaluation of general multi-center integrals using the expansion formula has been discussed.

1. INTRODUCTION

The accurate evaluation of multi-center integrals is essential for studying various electronic properties of molecules, complexes, cluster of atoms, solids and surfaces. Frequently, one uses linear combination of either the Gaussian-type orbitals or the Slater-type orbitals (STO) for performing multi-center calculations. Though the Gaussian-type orbitals are computationally convenient, they are inappropriate because of their poor radial dependence.

For the STO's several methods of attacks are known for the reduction of multi-center integrals each having its own merits and demerits. Recently, the present author[1] has been able to obtain a general and closed expression for the coefficients in the expansion of a general STO (including special cases). The importance of the derived expression for an STO lies in the fact that it contains terms only of the form $r^{k-\ell} \exp(\pm\eta r)$ where k is an integer >0; ℓ is the order of the coefficient and η is the exponent in the STO, which is very convenient and suitable for the analytical evaluation of the multi-center integrals. This has already been demonstrated by obtaining readily the general and closed form for the overlap matrix elements[1] between two STO's applicable for all quantum numbers involved.

C. A. Weatherford and H. W. Jones (eds.), ETO Multicenter Molecular Integrals, 35–51.

In the present paper we have shown that the same general expansion formula by the present author can be used to obtain the general and closed forms for the two-center hybrid and Coulomb integrals and the matrix elements involving hyperfine interaction operator, and the general expressions for the two-center exchange integrals and the three-center hyperfine matrix elements. The numerical results have also been given for cases involving s, p, d, f, g and h type STO's to prove the validity of the expressions. The discussion has also been given concerning the evaluation of the general multi-center matrix elements.

2. GENERAL EXPANSION FORMULA

Before presenting details for the derivation of the expressions for various integrals it is essential first to give the general and elegant mathematical form for the expansion of an STO from one center on to the other derived[1] previously as follows,

$$R^{N-1} e^{-\eta R} Y_L^M (\Theta \Phi) = \Sigma_\ell (\tfrac{1}{r}) \alpha_\ell (N\eta LM| a,r) Y_\ell^M(\theta,\phi) \quad (1)$$

where

$$\alpha_\ell (N\eta LM|ar) = \begin{cases} \Sigma_{k'=o}^{\ell+N} \left[A_{k'} (\tfrac{r}{a})^{k'-\ell} e^{-\eta r} + B_{k'} (\tfrac{r}{a})^{k'-\ell} e^{\eta r}\right], & r \leq a \\ \Sigma_{k'=o}^{\ell+N} C_{k'} (\tfrac{r}{a})^{k'-\ell} e^{-\eta r}, & r \geq a \end{cases} \quad (2)$$

where $A_{k'}$, $B_{k'}$, and $C_{k'}$ are independent of r and given by,

$$A_{k'} = -a^N e^{-\eta a} \Sigma_{k=o}^{k_{max}} (\tfrac{1}{\eta a})^{k+1} F_{k',k} (N\ell LM) \quad (3)$$

$$B_{k'} = -(-1)^{k'} A_{k'} \quad (4)$$

$$C_{k'} = A_{k'} + D_{k'} \quad (5)$$

$$D_{k'} = (-1)^{L+N-k'+1} a^N e^{\eta a} \Sigma_{k=o}^{k_{max}} (-\tfrac{1}{\eta a})^{k+1} F_{k',k}(N\ell LM) \quad (6)$$

In the above expressions the symbols used are the same as in Ref. (1). One notes that the left hand side of Eq. (1) is an STO located[2] at point B with the electron co-ordinates R, Θ, Φ whereas the right hand side is its expansion in terms of the electron co-ordinates r, θ, ϕ measured with respect to point A displaced from B by the distance a. The factors $F_{k',k}(N\ell LM)$ in Eqs. (3)-(6) depend on the quantum numbers $N\ell LM$ and are independent of the electron co-ordinates the exponent η or the distance a.

The above mathematical formula for the expansion of an STO has aroused enormous interest[3,4] in the literature and has been extremely useful for evaluating different kinds of integrals[5-7]. Making use of this formula a simple, closed and general expression for the overlap between two STO's has already been derived in Ref. (1). In the following we shall employ this formula to obtain expressions for hybrid, Coulomb and exchange integrals and the matrix elements valid for the general magnetic and electric hyperfine operators.

3. HYBRID INTEGRAL

The two-center hybrid integral has been defined as

$$I_h = (\Psi_{N_1L_1M_1\eta_1}(\vec{r}_1)\ \Psi_{N_2L_2M_2\eta_2}(\vec{r}_1);\Psi_{N_3L_3M_3\eta_3}(r_2)\ \Psi_{N_4L_4M_4\eta_4}(\vec{R}_2))$$

$$= \langle\Psi_{N_1L_1M_1\eta_1}(\vec{r}_1)\ \Psi_{N_3L_3M_3\eta_3}(\vec{r}_2)\left|\frac{1}{r_{12}}\right|\Psi_{N_2L_2M_2\eta_2}(\vec{r}_1)\ \Psi_{N_4L_4M_4\eta_4}(\vec{R}_2)\rangle \tag{7}$$

where $\Psi_{N_1L_1M_1\eta_1}(\vec{r}_1)$ is an STO (unnormalized) for an electron designated by its co-ordinate vector $\vec{r}_1$ with quantum numbers $N_1L_1M_1$ and the exponent η_1 with other symbols carrying similar meaning. The vector $\vec{R}_2$ stands for the co-ordinates of the other electron 2 measured from center B.

We next expand $\Psi_{N_4L_4M_4\eta_4}(\vec{R}_2)$ from center B on to the other center A by means of Eq. (1), integrate over angular co-ordinates of electrons 1 and 2 and simplify to obtain,

$$I_h = (-1)^{M_2}\ \delta_{M_1-M_2,M_4-M_3}\ {\sum_{\ell'=|L_1-L_2|}^{(L_1+L_2)}}'' \frac{4}{(2\ell'+1)}$$

$$ {\sum_{\ell=|L_3-\ell'|}^{(L_3+\ell')}}'' F_{L_3,\ell'}^{-M_3,M_3-M_4}(\ell)\ f_{\ell',\ell'+1}(N_1\eta_1N_2\eta_2|N_3\eta_3\alpha_\ell(N_4\eta_4L_4M_4)) \tag{8}$$

where "double primes" over the summations means that the summation index assumes the values with the step of 2, and, in general,

$$f_{\ell',\ell'+1}\,(N_1\eta_1N_2\eta_2|N_3\eta_3\ \alpha_\ell\ (N_4\eta_4L_4M_4))$$

$$= f^{(1)}_{\ell',\ell'+1} + f^{(2)}_{\ell',\ell'+1} + f^{(3)}_{\ell',\ell'+1} + f^{(4)}_{\ell',\ell'+1} + f^{(5)}_{\ell',\ell'+1} \qquad (9)$$

with

$$f^{(1)} = a^{N_1+N_2+N_3+1}\sum_{k'=o}^{\ell+N_4} A_{k'}\,(N_4\eta_4\ L_4M_4a)\sum_{\nu=o}^{\nu_{max}=N_1+N_2-\ell'-1}\frac{\nu_{max}\,!}{(\nu_{max}-\nu)\,!}$$

$$x\left[\frac{1}{(\eta_1+\eta_2)a}\right]^{\nu+1}\frac{(N_1+N_2+N_3+k'-\ell-\nu-1)\,!}{\left[a(\eta_1+\eta_2+\eta_3+\eta_4)\right]^{N_1+N_2+N_3+k'-\ell-\nu}} \qquad (10)$$

$$f^{(2)}_{\ell',\ell'+1} = a^{N_1+N_2+N_3+1}\sum_{k'=o}^{\ell+N_4} B_{k'}\,(N_4\eta_4\ell L_4M_4a)\sum_{\nu=0}^{\nu_{max}=N_1+N_2-\ell'-1}\frac{\nu_{max}\,!}{(\nu_{max}-\nu)\,!}$$

$$\left[\frac{1}{a(\eta_1+\eta_2)}\right]^{\nu+1} g_{\nu'_{max}}\left[a(\eta_1+\eta_2+\eta_3-\eta_4)\right]\ ;\quad \nu'_{max} = N_1+N_2+N_3+k'-\ell-\nu-1 \qquad (11)$$

$$f^{(3)}_{\ell',\ell'+1} = e^{-(\eta_1+\eta_2+\eta_3+\eta_4)a}\ a^{N_1+N_2+N_3+1}\sum_{k'=o}^{\ell+N_4} D_{k'}\,(N_4\eta_4\ell L_4M_4a)$$

$$\sum_{\nu=o}^{\nu_{max}=N_1+N_2-\ell'-1}\frac{\nu_{max}\,!}{(\nu_{max}-\nu)\,!}\ \frac{1}{\left[a(\eta_1+\eta_2)\right]^{\nu+1}}\ h_{\nu'_{max}}(a(\eta_1+\eta_2+\eta_3+\eta_4))\,;$$

$$\nu'_{max} = N_1+N_2+N_3+k'-\ell-\nu-1 \qquad (12)$$

$$f^{(4)}_{\ell',\ell'+1} = a^{N_1+N_2+N_3+1}\sum_{k'=o}^{\ell+N_4}\Big\{A_{k'}\,(N_4\eta_4\ L_4M_4a)\,\frac{(N_1+N_2+\ell')!}{\left[a(\eta_1+\eta_2)\right]^{N_1+N_2+\ell'+1}}$$

$$E_{N_3+k'-\ell-\ell'-1}\,(0,1,a(\eta_3+\eta_4)) - (-1)^{k'}\,E_{N_3+k'-\ell-\ell'-1}\,(0,1,a(\eta_3-\eta_4))\Big\}$$

$$- \sum_{\nu=o}^{\nu_{max} = N_1+N_2+\ell'} \frac{\nu_{max}\,!}{(\nu_{max}-\nu)\,!} \frac{1}{\left[a(\eta_1+\eta_2)\right]^{\nu+1}}$$

$$\left\{ E_{N_1+N_2+N_3+k'-\ell-\nu-1}(0,1;a(\eta_1+\eta_2+\eta_3+\eta_4) - (-1)^{k'} E_{N_1+N_2+N_3+k'-\ell-\nu-1}(0,1;a(\eta_3+\eta_1+\eta_2-\eta_4) \right\} \tag{13}$$

$$f^{(5)}_{\ell',\ell'+1} = a^{(N_1+N_2+N_3+1)} \sum_{k'=o}^{\ell+N_4} C_{k'}(N_4\eta_4\ell L_4 M_4 a) \left[\frac{(N_1+N_2+\ell')\,!}{\left[a(\eta_1+\eta_2)\right]^{N_1+N_2+\ell'+1}} \right.$$

$$E_{N_3+k'-\ell-\ell'-1}(1,\infty;a(\eta_3+\eta_4)) - \sum_{\nu=o}^{\nu_{max} = N_1+N_2+\ell'} \frac{\nu_{max}\,!}{(\nu_{max}-\nu)\,!} \frac{1}{\left[a(\eta_2+\eta_2)\right]^{\nu+1}}$$

$$\left. E_{N_1+N_2+N_3+k'-\ell-\nu-1}(1,\infty;a(\eta_1+\eta_2+\eta_3+\eta_4)) \right] \tag{14}$$

The symbols $E_n(a,b;\alpha)$, in general, stand for

$$E_n(a,b;\alpha) = \int_a^b x^n e^{-\alpha x} dx \tag{15}$$

which are easy to evaluate analytically. Also, $F^{m_1,m_2}_{\ell_1,\ell_2}(\ell_3)$ are the Clebsch-Gordan Coefficients defined by

$$F^{m_1,m_2}_{\ell_1,\ell_2}(\ell_3) = \left[\frac{(2\ell_1+1)(2\ell_2+1)(2\ell_3+1)}{4\pi} \right]^{\frac{1}{2}} (-1)^{m_3} \begin{pmatrix} \ell_1 & \ell_2 & \ell_3 \\ m_1 & m_2 & m_3 \end{pmatrix} \begin{pmatrix} \ell_1 & \ell_2 & \ell_3 \\ o & o & o \end{pmatrix} \tag{16}$$

Earlier attempts[8-11] for evaluation of the hybrid integrals are based on the combinations of auxiliary functions or their integrals. Numerical integration techniques[12-14] have also been developed for their evaluation by various authors. The above formula (Eq. (9)) is a closed and general analytical form of the hybrid integral and can be evaluated easily. The singularities which are present in the various terms of the expressions (Eq. (8)-(14)) for the hybrid integral cancel yielding finite results. In fact, as it is detailed in Section 7, it is possible to by-pass the singularities so that the evaluation of I_n becomes extremely easy.

4. COULOMB INTEGRAL

The two-center Coulomb integral has been defined by

$$I_c = \langle \Psi_{N_1L_1M_1\eta_1}(\vec{r}_1)\Psi_{N_3L_3M_3\eta_3}(\vec{R}_2) \left| \frac{1}{r_{12}} \right| \Psi_{N_2L_2M_2\eta_2}(\vec{r}_1)\Psi_{N_4L_4M_4\eta_4}(\vec{R}_2)\rangle \tag{17}$$

where all symbols carry the same meaning as above.

Combining first the angular parts of the STO's $\Psi_{N_3L_3M_3\eta_3}(\vec{R}_2)$ and $\Psi_{N_4L_4M_4\eta_4}(\vec{R}_2)$ by means of the Clebsch-Gordan coefficients and then expanding the combined radial and angular part centered at origin B on the center at origin A, one obtains, straightforwardly,

$$I_c = (-1)^{M_2+M_3} \left\{ \sum_{\ell'=|L_1-L_2|}^{L_1+L_2}\left(\frac{4\pi}{2\ell'+1}\right) F_{L_1,L_2}^{-M_1,M_2}(\ell') \sum_{\ell=|L_3-L_4|}^{L_3+L_4} F_{L_3,L_4}^{-M_3,M_4}(\ell) \right.$$

$$\left. f_{\ell',\ell'+1}(N_1\eta_1N_2\eta_2|1,o,\alpha_\ell'(N_4'\eta_4'\ell M_4'a)) \right\} \delta_{M_1-M_2,\,-M_3+M_4} \tag{18}$$

where $f_{\ell',\ell'+1}$ is the same as defined in Eqs. (9)-(14), and

$$N_4' = N_3+N_4-1$$

$$\eta_3' = \eta_3+\eta_4$$

$$M_4' = -M_3+M_4 \tag{19}$$

Remarkably, in the above expression for the Coulomb integral the factor $f_{\ell',\ell'+1}$ is the particular case of the general form given in Eqs. (9) to (14) occuring in hybrid integrals. Owing to this one does not require to expand extra labor for its evaluation by means of computers.

Particular formulas for the two-center Coulomb integrals employing STO's have been obtained much earlier by Mulliken et al.[15] and Roothaan[16]. However, these formulas are not in much use because of the inherent cancellation errors[17] occuring in some regions of parameters in calculations. Similar problems arise also in the recent formula given by Filter and Steinborn[18] as indicated by AntoLovic and Delhalle[19].

5. EXCHANGE INTEGRAL

The two-center exchange integral is given by

$$I_{ex} = \langle \Psi_{N_1L_1M_1\eta_1}(\vec{r}_1)\Psi_{N_3L_3M_3\eta_3}(\vec{R}_2)\left|\frac{1}{r_{12}}\right|\Psi_{N_2L_2M_2\eta_2}(\vec{R}_1)\Psi_{N_4L_4M_4\eta_4}(\vec{r}_2)\rangle \tag{20}$$

where the various symbols have been defined above. The angular integrations and simplifications yield,

$$I_{ex} = \delta_{M_2-M_1,M_3-M_4} \Sigma_{\ell'=o}^{\infty} \frac{4\pi}{(2\ell'+1)} \Sigma''^{L_1+\ell'}_{\ell_1=|L_1-\ell'|} F^{-M_1,-m'}_{L_1,\ell'}(\ell_1)$$

$$\Sigma''^{L_4+\ell'}_{\ell_2=|L_4-\ell'|} F^{m',M_4}_{\ell',L_4}(\ell_2)\; f_{\ell',\ell'+1}(N_1\eta_1\alpha_{\ell_1}(N_2\eta_2L_2M_2|a)|$$

$$\alpha_{\ell_2}(N_3\eta_3L_3M_3|a)N_4\eta_4\rangle \tag{21}$$

where,

$$f_{\ell',\ell'+1} = f^{(1)}_{\ell',\ell'+1} + f^{(2)}_{\ell',\ell'+1} + f^{(3)}_{\ell',\ell'+1} + f^{(4)}_{\ell',\ell'+1} \tag{22}$$

with

$$f^{(1)}_{\ell',\ell'+1} = f^{(1)}_{\ell',\ell'+1}(N_1\eta_1\alpha_{\ell_1}(N_2\eta_2L_2M_2|ar_1)|\alpha_{\ell_2}(N_3\eta_3L_3M_3|ar_2)N_4\eta_4)$$

$$= a^{N_1+N_4+1} \Sigma^{\ell_1+N_2}_{k_1'=o} \Sigma^{\ell_2+N_3}_{k_2'} \Big[A_{k_2'}(N_3\eta_3\ell_2L_3M_3a)\{A_{k_1'}(N_2\eta_2\ell_1L_2M_2a)$$

$$\left(\frac{\nu_{max}!}{a(\eta_1+\eta_2)^{\nu_{max}+1}} E_{N_4+k_2'-\ell_2-\ell'-1}(o,1;a(\eta_3+\eta_4)) - \Sigma^{\nu_{max}}_{\nu=o} \frac{\nu_{max}!}{(\nu_{max}-\nu)!} \right.$$

$$\frac{1}{\left[a(\eta_1+\eta_2)\right]^{\nu+1}} \; E_{N_4+k_2'-\ell_2+N_1+k_1'-\ell_1-\nu-1}(o,1;a(\eta_1+\eta_2+\eta_3+\eta_4))\Bigg) + B_{k_1'}(N_2\eta_2\ell_1\, L_2M_2a)$$

$$\left(\frac{(\nu_{max})\,!}{\left[a(\eta_1-\eta_2)\right]^{(\nu_{max}+1)}} E_{N_4-\ell'-1+k_2'-\ell_2}(o,1,a(\eta_3+\eta_4)) - \Sigma_{\nu=o}^{\nu_{max}} \frac{\nu_{max}\,!}{(\nu_{max}-\nu)\,!} \frac{1}{\left[a(\eta_1-\eta_2)\right]^{\nu+1}}\right.$$

$$\left. E_{N_4+k_2'-\ell_2+N_1+k_1'-\ell_1-\nu-1}(0,1;a(\eta_3+\eta_4+\eta_1-\eta_2)\right) \Big\}$$

$$+ B_{k_2'}(N_3\eta_3\ell_2L_3M_3a) \quad \{\, \eta_3 \rightarrow (-\eta_3)\, \}\Big] \tag{23}$$

where

$$\nu_{max} = N_1+\ell'+k_1'-\ell_1 \tag{24}$$

In the above expression $\{\eta_3 \rightarrow (-\eta_3)\}$ stands for the terms contained in the preceeding curly bracket with η_3 replaced by $(-\eta_3)$. The term $f^{(2)}_{\ell',\ell'+1}$ in Eq. (22) is

$$f^{(2)}_{\ell',\ell'+1}(N_1\eta_1\alpha_{\ell_1}(N_2\eta_2L_2M_2|ar_1)|\alpha_{\ell_2}(N_3\eta_3L_3M_3|ar_2)N_4\eta_4)$$

$$= a^{N_1+N_4+1} \; \Sigma_{k'=o}^{\ell_2+N_3} C_{k_2'}(N_3\eta_3\ell_2L_3M_3) \Big[E_{N_4+k_2'-\ell_2-\ell'-1}(1,\infty;a(\eta_3+\eta_4))$$

$$\Sigma_{k_1'=o}^{\ell_1+N_2} \{A_{k_1'}(N_2\eta_2\ell_1L_2M_2)\, g_{N_1\ell_1+k_1'-\ell_1}(a(\eta_1+\eta_2)) + B_{k_1'}(N_2\eta_2\ell_1L_2M_2)$$

$$g_{N_1+\ell'+k_1'-\ell_1}(a(\eta_1-\eta_2)) + C_{k_1'}(N_2\eta_2\ell_1L_2M_2)\, e^{-a(\eta_1+\eta_2)} h_{N_1+\ell'+k'-\ell_1}$$

$$(a(\eta_1+\eta_2))\} - \Sigma_{k_1'}^{\ell_1+N_2} C_{k_1'=o}(N_2\;2\;1L_2M_2)\; \Sigma_{\nu=o}^{\nu_{max}} \frac{\nu_{max}\,!}{(\nu_{max}-\nu)!} \frac{1}{\left[a(\eta_1+\eta_2)\right]^{\nu+1}}$$

$$E_{N_4+k_2'+N_1+k_1'-\ell_2-\ell_1-\nu-1}(1,\infty;a(\eta_1+\eta_2+\eta_3+\eta_4))\Big] \tag{25}$$

The terms $f^{(3)}_{\ell',\ell'+1}$ and $f^{(4)}_{\ell',\ell'+1}$ in Eq. (22) are related to $f^{(1)}_{\ell',\ell'+1}$ and $f^{(2)}_{\ell',\ell'+1}$, respectively, in the following manner,

$$f^{(3)}_{\ell',\ell'+1}(N_1\eta_1\alpha_{\ell_1}(N_2\eta_2L_2M_2|ar_1)|\alpha_{\ell_2}(N_3\eta_3L_3M_3|ar_2)N_4\eta_4)$$

$$= f^{(1)}_{\ell',\ell'+1}(N_4\eta_4\alpha_{\ell_2}(N_3\eta_3L_3M_3|ar_1)|\alpha_{\ell_1}(N_2\eta_2L_2M_2|ar_2)N_1\eta_1) \quad (26)$$

$$f^{(4)}_{\ell',\ell'+1}(N_1\eta_1\alpha_{\ell_1}(N_2\eta_2L_2M_2|ar_1)\alpha_{\ell_2}(N_3\eta_3L_3M_3|ar_2)N_4\eta_4)$$

$$= f^{(2)}_{\ell',\ell'+1}(N_4\eta_4\alpha_{\ell_2}(N_3\eta_3L_3M_3|ar_1)|\alpha_{\ell_1}(N_2\eta_2L_2M_2|ar_2)N_1\eta_1) \quad (27)$$

5. MATRIX ELEMENT FOR MAGNETIC AND ELECTRIC HYPERFINE INTERACTIONS

The two-center matrix element for the hyperfine interaction is

$$I = \langle N_1L_1M_1\eta_1 \left| \frac{Y^m_\ell\,(\Theta\ \Phi)}{R^n} \right| N_2L_2M_2\eta_2 \rangle \quad (28)$$

where the wavefunctions are centered at one location (Say A) and the operator at the other location, (Say B). There are two options for evaluating these integrals, by expanding the operator on to the center A or by expanding the combination of the wavefunctions from center A on to center B. If we follow the later procedure, we have

$$I = \sum_{\ell=|L_1-L_2|}^{''(L_1+L_2)} F^{-M_1,M_2}_{L_1,L_2}(\ell)\ S_{(-n+1)\ell m(\eta=o);\ (N_1+N_2-1),\ell,(M_2-M_1);\ (\eta_1+\eta_2)} \quad (29)$$

where $S_{N_1L_1M_1\eta_1;N_2,L_2,M_2,\eta_2}$ is the overlap-type integral the evaluation of which has been explained previously in Ref.(1).

Alternatively, if one expands the operator $Y^m_\ell(\theta,\phi)/r^3$ instead of the wavefunctions, one gets easily the general as well as the particular integrals as have been shown previously[5]. In this case one may require to consider also the contributions arising from the δ-function appearing in the expansion of the operator[20].

6. THREE-CENTER MAGNETIC AND ELECTRIC HYPERFINE INTERACTION MATRIX ELEMENTS

The matrix element of interest is,

$$I_{hf} = \langle \Psi_{N_1L_1M_1\eta_1}(\vec{R}) \left| \frac{Y^m_\ell\ (\theta,\phi)}{r^n} \right| \Psi_{N_2L_2M_2\eta_2}(\vec{R}')\rangle \quad (30)$$

where R_1 and R_1' stand for the co-ordniates of an electron with respect to center A or B, respectively. r,θ,ϕ are the co-ordinates of the electron measured with respect to the origin at a general point C. For simplicity we assume that the origin C and the centers A and B are co-linear. The most general case of nonco-linear A,B,C can be traced analogously by using appropriate rotational groups.

In the present case we expand the STO's $\Psi_{N_1L_1M_1\eta_1}(\vec{R})$ and $\Psi_{N_2L_2M_2\eta_2}(\vec{R}')$ on to the origin with the help of the expressions given in Eqs. (1)-(6). Then, the integration over the angular co-ordinates and simplification yield,

$$I_{hf} = \delta_{M_1,m+M_2} \sum_{\ell_1} \sum_{\ell_2=|\ell\ \ell_1|}^{\ell+\ell_1} F^{m,M_2}_{\ell,\ell_2}(\ell_1) \sum_{k_1'=o}^{\ell_1+N_1} \sum_{k_2'=o}^{\ell_2+N_2} \left(\frac{1}{a_1}\right)^{k_1'-\ell_1} \left(\frac{1}{a_2}\right)^{k_2'-\ell_2} \Big[A^{(1)}_{k_1'} A^{(2)}_{k_2'} a_1^{n'+1}$$

$$E_{n'}(0,\infty;a_1(\eta_1+\eta_2)) + A^{(1)}_{k_1'} B^{(2)}_{k_2'} a_2^{n'+1} E_{n'}(0,1;a_2(\eta_1-\eta_2))$$

$$+B^{(1)}_{k_1'} A^{(2)}_{k_2'} a_1^{n'+1} E_{n'}(o,1,a_1(\eta_2-\eta_1)) + B^{(1)}_{k_1'} B^{(2)}_{k_2'} a_1^{n'+1}$$

$$E_{n'}(o,1;(-\eta_1-\eta_2)a_1) + D^{(1)}_{k_1'} A^{(2)}_{k_2'} a_1^{n'+1} E_{n'}(1,\infty;a_1(\eta_1+\eta_2))$$

$$+D^{(1)}_{k_1'} B^{(2)}_{k_2'} a_1^{n'+1} E_{n'}(1, \frac{a_2}{a_1}; a_1(\eta_1-\eta_2)) + (A^{(1)}_{k_1'} + D^{(1)}_{k_1'}) D^{(2)}_{k_2'}$$

$$a_2^{n'+1} E_{n'}(1,\infty;a_2(\eta_1+\eta_2)) \Big] \qquad (31)$$

where $n'=k_1'-\ell_1+k_2'-\ell_2-n$ and a_1 and a_2 are the distances of the centers A and B from the origin C. In the above expression the superscript "(1)" on the symbols $A^{(1)}_{k_1'}$, $B^{(1)}_{k_1'}$, $C_{k_1'}$ and $D^{(1)}_{k_1'}$ means that these parameters correspond to the expansion of the STO $\Psi_{N_1L_1M_1\eta_1}$; similarly, the subscript "(2)" on $A^{(2)}_{k_2'}$, $B^{(2)}_{k_2'}$, $C^{(2)}_{k_2'}$, $D^{(2)}_{k_2'}$ indicate that these correspond to the expansion of the STO $\Psi_{N_2L_2M_2\eta_2}$. An expression similar to Eq.(31) for n=3, ℓ=2, m=o has been derived previously[5].

7. ELIMINATION OF SINGULARITIES AND CALCULATIONS

The exponential integrals $E_n(a,b;\alpha)$ defined in Eq. (15) and occurring in various multi-center integrals is singular when a=o and n is negative. In that case it is possible to by-pass the singularities by making use of the following expressions for the exponential integrals throughout in the multi-center integrals,

$$E_{n'}(o,1;\mp\alpha) = \int_o^1 \frac{e^{\pm\alpha t}}{t^{n'}}$$

$$\equiv \frac{(\pm\alpha)^{n'-1}}{(n'-1)\,!}\Big[\Psi(n')-\ell n(|\alpha|)\Big] - \int_1^{\mp\infty} \frac{e^{\pm\alpha t}}{t^{n'}}\,dt \qquad (32)$$

where n'>o and α>o and $\Psi(n)$ is the logarithmic derivative of the Gamma functions. The $E_{n'}$ integral can be evaluated as accurately as required by an extenstion of the method described previously[21]. Other special cases of Eq. (15) are amenable to the methods well-known in the literature[1] and are easily evaluated.

In order to prove the validity of the above expressions, we present the calculated values of the integrals in Tables I-III for various quantum numbers involving s,p,d,f,g,h orbitals for two-center hybrid and Coulomb integrals and two-center and three-center hyperfine interaction matrix elements. For the hybrid (Table I) and Coulomb (Table II) integrals we have chosen η_1=0.5, η_2=1.0; η_3=1.5; η_4=2.0, a=2.0 and the STO's are unnormalized. For the hyperfine matrix elements we have taken the following orbitals appropriate for the evaluation of nuclear quadrupole interactions associated with Fe^{57m} nucleus in hemoglobin hydroxide[5,6] and hemoglobin cyanide[22]:

$$(1s) = \sqrt{\frac{Z_1^3}{\pi}}\ \exp(-Z_1 R)$$

$$(2s) = \sqrt{\frac{Z_2^5}{3\pi}}\ R\exp(-Z_2 R)$$

$$(2p_o) = \sqrt{\frac{4}{3} Z_2^5}\ R\exp(-Z_2 R)\ Y_1^0(\Theta,\Phi)$$

$$(2p_{\pm1}) = \sqrt{\frac{4}{3} Z_2^5}\ R\exp(-Z_2 R)\ Y_1^{\pm1}(\Theta,\Phi) \qquad (33)$$

TABLE I. List of two-center hybrid integrals between STO's for $\eta_1=0.5$; $\eta_2=1.0$; $\eta_3=1.5$; $\eta_4=2.0$; a=2.0

$N_1L_1M_1$	$N_2L_2M_2$	$N_3L_3M_3$	$N_4L_4M_4$	I_h
1s	1s	1s	$4f_o$	$-0.136202\text{x}10^{-1}$
$2p_o$	$2p_o$	1s	$4f_o$	$-0.401505\text{x}10^{-1}$
1s	1s	$4f_o$	$4f_o$	$-0.226192\text{x}10^{-1}$
1s	1s	1s	$3d_o$	$0.896274\text{x}10^{-2}$
1s	1s	$3d_o$	$3d_o$	$-0.114136\text{x}10^{-2}$
1s	1s	$6h_o$	$3d_o$	0.162865
$2p_o$	$2p_o$	$4f_o$	$4f_o$	$-0.826819\text{x}10^{-1}$
$2p_o$	$2p_o$	$2p_o$	$3d_o$	$-0.381092\text{x}10^{-2}$
$2p_o$	$2p_o$	$3d_o$	$3d_o$	$-0.322357\text{x}10^{-2}$
$2p_o$	$2p_o$	$6h_o$	$3d_o$	0.918496
1s	1s	$6h_o$	$4f_o$	0.336383
1s	1s	1s	1s	$0.377882\text{x}10^{-2}$
1s	1s	1s	$2p_o$	$-0.608120\text{x}10^{-2}$
1s	1s	$2p_o$	1s	$0.155003\text{x}10^{-2}$
$2p_{+1}$	1s	1s	$2p_{+1}$	$0.101642\text{x}10^{-2}$
$2p_{\pm1}$	$2p_{\pm1}$	$2p_{\pm1}$	$2p_{\pm1}$	$0.178288\text{x}10^{-1}$
$2p_{-1}$	$2p_{+1}$	$2p_{+1}$	$2p_{-1}$	$0.134955\text{x}10^{-2}$

TABLE II. Tabulation of the two-center Coulomb integrals between STO's for $\eta_1=0.5$; $\eta_2=1.0$; $\eta_3=1.5$; $\eta_4=2.0$; a=2.0.

$N_1L_1M_1$	$N_2L_2M_2$	$N_3L_3M_3$	$N_4L_4M_4$	I_c
1s	1s	1s	1s	0.117001×10^{-1}
1s	1s	1s	$2p_o$	-0.179928×10^{-2}
1s	1s	$2p_o$	1s	-0.179928×10^{-2}
1s	$2p_o$	1s	1s	0.752922×10^{-2}
$2p_o$	1s	1s	1s	0.752922×10^{-2}
$2p_{+1}$	1s	1s	$2p_{+1}$	0.198491×10^{-2}
1s	$2p_{+1}$	$2p_{+1}$	1s	0.198491×10^{-2}
$2p_{+1}$	$2p_{+1}$	$2p_{+1}$	$2p_{+1}$	0.425312×10^{-1}
$2p_{-1}$	$2p_{-1}$	$2p_{-1}$	$2p_{-1}$	0.425312×10^{-1}
$2p_{-1}$	$2p_1$	$2p_{+1}$	$2p_{-1}$	0.212319×10^{-2}
1s	1s	1s	$4f_o$	-0.419785×10^{-3}
$2p_o$	$2p_o$	1s	$4f_o$	-0.531931×10^{-3}
1s	1s	$2p_o$	$3d_o$	-0.359756×10^{-2}
1s	1s	1s	$3d_o$	0.690776×10^{-3}
1s	1s	$3d_o$	$2p_o$	-0.359756×10^{-2}
$4f_o$	1s	$2p_o$	$2p_o$	0.870596×10^{-2}
$2p_o$	$2p_o$	$4f_o$	$4f_o$	0.474823×10^{0}
$2p_o$	$2p_o$	$2p_o$	$3d_o$	-0.412145×10^{-2}
$2p_o$	$2p_o$	$3d_o$	$3d_o$	0.113441
$5g_o$	1s	$2p_o$	$2p_o$	0.165590×10^{-1}
$4f_o$	1s	1s	$2p_o$	0.445948×10^{-2}
$5g_o$	1s	$2p_o$	1s	0.989952×10^{-2}

TABLE III. List of two- and three-center hyperfine interaction integrals useful for evaluating nuclear quadrupole interactions associated with Fe^{57m} in hemoglobin hydroxide and hemoglobin cyanide (without δ-function term).

Matrix elements	$(1/\sqrt{4\pi})$ a.u.
$<(1S)_O \| Y_2^0/r^3 \| (1S)_O>$	5.3026×10^{-2}
$<(2S)_O \| Y_2^0/r^3 \| (2S)_O>$	5.0112×10^{-2}
$<(2p_\sigma)_O \| Y_2^0/r^3 \| (2p_\sigma)_O>$	6.8160×10^{-2}
$<(1S)_O \| Y_2^0/r^3 \| (2S)_O>$	2.0×10^{-7}
$<(1S)_O \| Y_2^0/r^3 \| (2p_\sigma)_O>$	2.4701×10^{-3}
$<(2S)_O' \| Y_2^0/r^3 \| (2p_\sigma)_O>$	2.8374×10^{-2}
$<(2p_{+1})_O \| Y_2^0/r^3 \| (2p_{+1})_O>$	4.5425×10^{-2}
$<(2S)_H \| Y_2^0/r^3 \| (1S)_H>$	1.5160×10^{-2}
$<(2S)_O \| Y_2^0/r^3 \| (1S)_H>$	9.2335×10^{-4}
$<(2p_\sigma)_O \| Y_2^0/r^3 \| (1S)_H>$	5.6300×10^{-5}
$<(1S)_O \| Y_2^0/r^3 \| (1S)_H>$	3.5759×10^{-5}
$<(2S)_C \| Y_2^0/r^3 \| (2S)_C>$	5.2479×10^{-2}
$<(2p_\sigma)_C \| Y_2^0/r^3 \| (2p_\sigma)_C>$	8.2077×10^{-2}
$<(2p_\pi)_C \| Y_2^0/r^3 \| (2p_\pi)_C>$	3.6907×10^{-2}
$<(2S)_N \| Y_2^0/r^3 \| (2S)_N>$	1.2308×10^{-2}

TABLE III - (continued)

Matrix elements	$(1/\sqrt{4\pi})$ a.u.
$<(2p_\sigma)_N \vert Y_2^o/r^3 \vert (2p_\sigma)_N>$	1.4190×10^{-2}
$<(2p_\pi)_N \vert Y_2^o/r^3 \vert (2p_\pi)_N>$	1.0430×10^{-2}
$<(2S\ )_C \vert Y_2^o/r^3 \vert (2p_\sigma)_C>$	4.0228×10^{-2}
$<(2S\ )_N \vert Y_2^o/r^3 \vert (2p_\sigma)_N>$	4.8999×10^{-3}
$<(2S\ )_C \vert Y_2^o/r^3 \vert (2S)_N>$	1.01221×10^{-2}
$<(2S\ )_C \vert Y_2^o/r^3 \vert (2p_\sigma)_N>$	-1.24542×10^{-2}
$<(2p_\sigma)_C \vert Y_2^o/r^3 \vert (2S)_N>$	5.43696×10^{-3}
$<(2p_\sigma)_C \vert Y_2^o/r^3 \vert (2p_\sigma)_N$	-3.12315×10^{-3}
$<(2p_\pi)_C \vert Y_2^o/r^3 \vert (2p_\pi)_N$	4.44106×10^{-3}

with Z_1=7.70 and Z_2=2.275 for oxygen, Z_1=1.601 and Z_2=1.561 for C and Z=1.924 and Z_2=1.917 for N; for the hydrogen orbital we have assumed the STO as

$$(1s) = \frac{1}{\sqrt{\pi}} \exp(-R) \tag{34}$$

8. CONCLUSION

From our general expression of Ref.(1) for the expansion of an STO from one center onto the other, we have derived various two-and three-center integrals which turn out to be simple for evaluation (see Tables I-III). One may use similar technique to evaluate general four-center integrals for which the details will be presented elsewhere [23]. For four-center integrals one may encounter problems associated with the convergence and cancellation errors which could be surmounted by judicial arrangement of terms and, in worst cases, by making use of prime-number algebra and long precision calculations.

REFERENCES

1. Sharma, R.R.: 1981, Physical Rev. A 13, p.517.

2. Sharma, R.R.: 1968, J. Math. Phys. 9, p.505.

3. Jones, H.W.: 1980, Int. J. Quantum Chem. 18, p.709; 1981, 19, p.567.

4. Viccaro, M.H. de A.: 1976, Int. J. Quantum Chem. 10, p.1075; Silverstone, H.J. and Moats, P.K.: 1977, Phys. Rev. A 16, p.173; Rashid, M.A.: (to appear), J. Math. Phys.

5. Sharma, R.R.: 1978, Phys. Rev. 18, p.726; Sharma, R.R.: 1972, Phys. Rev. B 6, p.4310.

6. Sharma, R.R. and Kolman, R.: 1981, J. Chem. Phys. 68, p.2516.

7. Viccaro, M.H. de A., Sundaram, S. and Sharma, R.R.: 1981, Bull. Am. Phys. Soc. 26, p.287; 1981, 25, p.326.

8. Barnett, M.P. and Coulson, C.A.: 1951, Phil. Tran. A 243, p.221.

9. Ruedenberg, K., Roothaan, C.C.J. and Jaunzemis, W.: 1965, J. Chem. Phys. 24, p.201.

10. Kopineck, H.: 1951, Z. Naturforsch 6a, p.177.

11. Kotani, M., Amemiya, A., Ishigura, E. and Kimura, T.: 1963, Tables of Molecular Integrals; Maruzen, Tokyo.

12. Wahl, A.C., Cade, P.E. and Roothaan, C.C.J.: 1964, J. Chem. Phys. 41, p.2578.

13. Chirstoffersen, R.E. and Ruedenberg, K.: 1968, J. Chem. Phys. 49, p.4285.

14. Milleur, M.B. and Matcha, R.L.: 1972, J. Chem. Phys. 57, p.3029.

15. Mulliken, R., Rieke, C., Orloff, D. and Orloff, H.: 1949, J. Chem. Phys. 12, p.1278.

16. Roothaan, C.C.J.: 1951, J. Chem. Phys. 19, p.1445.

17. Harris, F.E.: 1969, J. Chem. Phys. 51, p.4708.

18. Filter, E. and Steinborn, E.O.: 1978, Phys. Rev. A 18, p.1.

19. Antolovic, D. and Delhalle, J.: 1980, Phys. Rev. A 21, p.1815.

20. Pitzer, R.M., Kern, C.W. and Lipscomb, W.N.: 1962, J. Chem. Phys. 37, p.267.

21. Sharma, R.R. and Zahuri, B.: 1977, J. Comp. Phys. 25, p.199.

22. Sharma, R.R.: 1979, Bull. Am. Phys. Soc. 24, p.319.

23. Sharma, R.R.: (to be published).

THE PHILOSOPHY AND STRATEGY FOR STO INTEGRAL EVALUATION USING THE C-MATRIX ALGEBRAIC SINGLE-CENTER EXPANSION METHOD

Herbert W. Jones
Institute for Molecular Computations and Physics Department
Florida A&M University, Tallahassee, Florida 32307

Sufficient work has now been done to establish the viability of the algebraic C-matrix single-center expansion method for the evaluation of integrals over Slater-type orbitals. Explicit formulas for two-center overlap[1,2], Coulomb[3], and hybrid[4] integrals have been automatically generated by computer using this method. Accurate numerical evaluation [5] of these formulas for all values of parameters can be achieved by machine manipulation of these formulas into appropriate Taylor series. The policy of carrying out all algebraic manipulations by computer and using different formulas for different ranges of parameters re-establishes the use of formulas for integral evaluation after the abandonment of this approach many years ago.

It is our insistence on formulas, as represented by coefficient arrays, that distinguishes our single-center method of doing integrals from that of Harris and Michels[6], and that of Sharma[7]. Computer generated formulas present a much more penetrating insight into the mathematical nature of a problem than does pure "number crunching". Because all previous single-center methods give poor convergence when displacement distances are large, it appears that the only hope for the single-center method in this general case is to generate formulas for integrals.

We make a conjecture that all STO molecular integrals can be expressed in closed form, i.e., as a sum of polynomials multiplied by well-known functions. We further speculate that the appropriate parameters of such a formula for three-center integrals would be the three sides of the triangle formed by connecting the three centers, and for four-center integrals the six sides of the pyramid-like figure formed by the lines connecting the four centers. This is made plausible by the fact that Legendre polynomials can be expressed as a sum of the powers of the cosine of an angle, and the cosine can be expressed in terms of sides of a triangle.

To illustrate how a formal infinite series may be terminated in a low order polynomial we will take the well-known case of the overlap

C. A. Weatherford and H. W. Jones (eds.), ETO Multicenter Molecular Integrals, 53–59.

integral between two normalized 1s orbitals which have screening constants set to one (ζ=1).

Let one orbital χ_a be centered at a distance a along the negative z axis and let the other orbital χ_b be centered a distance a along the positive z axis. The overlap integral S is given by

$$S = \int \chi_a \chi_b \, dv$$

Expanding the orbitals in spherical harmonics about the origin,

$$\chi_a = A\,K \sum_{\ell}^{\infty} k_\ell (-1)^\ell \, \alpha_\ell \, Y_\ell^M (\theta,\phi)$$

$$\chi_b = A\,K \sum_{\lambda}^{\infty} k_\lambda \, \alpha_\lambda \, Y_\lambda^M (\theta,\phi)$$

where $A = (2\zeta)^{N+\frac{1}{2}} \left[(2N)!\right]^{-\frac{1}{2}}$

$$K = \frac{1}{\zeta^{N-1}} \left[\frac{(2L+1)\,(L-M)!}{(L+M)!}\right]^{\frac{1}{2}} (-1)^M,$$

and

$$k_\ell = \left[\frac{(\ell+M)!}{(2\ell+1)\,(\ell-M)!}\right]^{\frac{1}{2}}$$

Hence,

$$S = A^2K^2 \sum_\ell \sum_\lambda k_\ell\, k_\lambda (-1)^\ell \int r^2 dr\, \alpha_\ell \alpha_\lambda \int d\omega\, Y_\ell^M(\theta,\phi)\, Y_\lambda^M(\theta,\phi)$$

Invoking the ortho-normal properties of spherical harmonics we have

$$S = A^2K^2 \sum_\ell^\infty k_\ell^2 \,(-1)^\ell \int r^2 dr\, \alpha_\ell^2 = \sum_{\ell=o}^\infty S_\ell$$

with

$$S_\ell = A^2K^2k_\ell^2 \,(-1)^\ell \int r^2 dr\, \alpha_\ell^2$$

Now,

$$\alpha_\ell^{NLM} = h_\ell \sum_{i=o}^{N+L+\ell} \sum_{j=o}^{N+\ell} C_\ell^{NLM}(i,j) . H_{ij} . (\zeta a)^{i-L-\ell-1} \, (\zeta r)^{j-\ell-1}$$

where $h_\ell = \dfrac{(2\ell+1)(\ell-M)!}{2(\ell+M)!}$

and $$H_{ij} = \begin{cases} e^{-\zeta a}\left[(-1)^j e^{\zeta r} - e^{-\zeta r}\right] & r<a, \\ e^{-\zeta r}\left[(-1)^i e^{\zeta a} - e^{-\zeta a}\right] & r>a \end{cases}$$

Setting N=1, L=0, M=0, and $\zeta=1$

$$S_\ell = \sum_i \sum_j \sum_p \sum_q (-1)^\ell (2\ell+1)\, C_\ell^{100}(i,j)\, C_\ell^{100}(p,q)\, a^{i+p-2\ell-2}$$

$$\times \int_0^\infty dr\, H_{ij}\, H_{pq}\, r^{j+q-2\ell}$$

The results of carrying out this integral for $\ell=0,1,2$, and 3 are shown in Table I. (For clarity, we use decimals instead of the more desirable exact fractions. Also, a and its power has been subsequently typed in beside each coefficient value). Now we come to a difficult part of our procedure: we must guess the final form of our formula. Let us assume that the final formula will be $\exp(-2a)$ multiplied by a polynomial. Since $\exp(-2a) \times \exp(2a) = 1$, we may factor out $\exp(-2a)$ from the terms without an exponential factor and do a multiplication by $\exp(2a)$. $\mathrm{Exp}(2a)$ is expanded in a power series with sufficient terms for each multiplication so that common powers have matching coefficients. As expected, all coefficients of inverse powers of a add to zero since the overlap integral must remain finite as a goes to zero. Table II shows the results of these operations. Note the occurrence of zero coefficients for higher ℓ values. Table III shows the results of adding one expanded S_ℓ expression from Table II at a time. With the addition of only S_o and S_1 the true overlap formula is developed if we only consider the first three terms of the infinite series. The addition of S_2 and S_3 only cause the coefficients of higher powers of a to become zero or approach zero. This would appear to be the result of adding all higher S_ℓ expressions. Thus we may say that we have "projected to infinity" and terminated the infinite series into a three term polynomial.

It has been possible[8] to derive Sugiura's[9] formula for the two-center exchange integral for 1s orbitals with equal screening constants by this method. Whether or not it will be possible to "project to infinity" for every kind of molecular integral remains to be seen.

Table I. Closed partial sum formulas for S_0, S_1, S_2 and S_3 generated by computer. The powers of a have been typed-in for easy identification of the formula coefficients.

$$S_o = \frac{2.5}{a^2} + e^{-2a}\left(-\frac{5.0}{a} - \frac{2.5}{a^2} - 4.0 - 1.3a\right)$$

$$S_1 = \frac{-7.5}{a^2} + \frac{52.5}{a^4} + e^{-2a}\left(\frac{-55.0}{a} - \frac{97.5}{a^2} - \frac{105.0}{a^3} - \frac{52.5}{a^4} - 20.0 - 4.0a\right)$$

$$S_2 = \frac{12.5}{a^2} - \frac{262.5}{a^4} + \frac{2362.5}{a^6} + e^{-2a}\left(\frac{-305.0}{a} - \frac{1062.5}{a^2} - \frac{2625.0}{a^3} - \frac{4462.5}{a^4} - \frac{4725.0}{a^5} - \frac{2362.5}{a^6} - 60.0 - 6.67a\right)$$

$$S_3 = \frac{-17.5}{a^2} + \frac{735.0}{a^4} - \frac{16537.5}{a^6} + \frac{181912.5}{a^8} + e^{-2a}\left(\frac{-1155.0}{a} - \frac{6597.5}{a^2} - \frac{27930.0}{a^3} - \frac{88935.0}{a^4} - \frac{209475.0}{a^5}\right.$$

$$\left. - \frac{347287.5}{a^6} - \frac{363825.0}{a^7} - \frac{181912.5}{a^8} - 140.0 - 9.33a\right)$$

Table II. Power series expansions for S_o, S_1, S_2 and S_3 obtained by factoring out exp(-2α) from the exact expressions. The factoring is accomplished by multiplying part of the exact expression by exp(2α).

$$S_o = e^{-2\alpha}\,(1.0000 + 2.0000\alpha + 1.6667\alpha^2 + .6667\alpha^3 + .2222\alpha^4 + .0635\alpha^5 + \ldots.)$$

$$S_1 = e^{-2\alpha}\,(.0000 + .0000\alpha - .3333\alpha^2 - .6667\alpha^3 - .3333\alpha^4 - .1164\alpha^5 + \ldots.)$$

$$S_2 = e^{-2\alpha}\,(.0000 + .0000\alpha + .0000\alpha^2 + .0000\alpha^3 + .1111\alpha^4 + .0683\alpha^5 + \ldots.)$$

$$S_3 = e^{-2\alpha}\,(.0000 + .0000\alpha + .0000\alpha^2 + .0000\alpha^3 + .0000\alpha^4 + .0166\alpha^5 + \ldots.)$$

Table III. The development of a terminated polynomial as part of the overlap formula, illustrated by the addition of partial sums S_o, S_1, S_2 and S_3.

$$S_o = e^{-2a}\,(1.0000 + 2.0000a + 1.6667a^2 + .6667a^3 + .2222a^4 + .0635a^5 + \ldots.)$$

$$S_o+S_1 = e^{-2a}\,(1.0000 + 2.0000a + 1.3333a^2 + .0000a^3 - .1111a^4 - .0529a^5 + \ldots.)$$

$$S_o+S_1+S_2 = e^{-2a}\,(1.0000 + 2.0000a + 1.3333a^2 + .0000a^3 + .0000a^4 + .0154a^5 + \ldots.)$$

$$S_o+S_1+S_2+S_3 = e^{-2a}\,(1.0000 + 2.0000a + 1.3333a^2 + .0000a^3 + .0000a^4 - .0012a^5 + \ldots.)$$

REFERENCES

1. Jones, H.W.: 1980, Int. J. Quantum Chem. 18, p.709.

2. Jones, H.W.: 1981, Int. J. Quantum Chem. 19, p.567.

3. Jones, H.W.: 1981, Int. J. Quantum Chem. 20.

4. Jones, H.W.: 1981, Int. J. Quantum Chem. Symp. 15.

5. Jones, H.W.: (submitted), Int. J. Quantum Chem.

6. Harris, F.E. and Michels, H.H.: 1965, J. Chem. Phys. 43, S165.

7. Sharma, R.R.: 1976, Phys. Rev. A13, p.517.

8. Jones, H.W.: 1978, Bull. Am. Phys. Soc. 23, p.562.

9. Sugiura, Y.: 1927, Z. Physik 45, p.484.

EXPANSION OF A FUNCTION ABOUT A DISPLACED CENTRE

M.A. Rashid**

International Centre for Theoretical Physics, Trieste, Italy.
**Permanent address: Department of Mathematics, Ahmadu Bello University, Zaria, Nigeria.

We review the progress recently made in obtaining closed form expressions for the expansion of general orbitals about a displaced centre and establish the equivalence between different expansions. We also examine how these expressions do have the desired limit as the displacement approaches zero.

1. INTRODUCTION

In the last few years, some progress has been made in obtaining closed form expressions for the expansion of a function about a displaced centre[1-3]. These expressions are extremely useful in the calculation of multi-centre integrals. The purpose of the present note is to review this progress outlining some of the problems that seem to arise in this connection.

We organize our work as follows.

In Section II we rederive the famous expansion formula for $R^L Y_L^M(\Theta,\Phi)$ [4] and point out that the limit when the displacement vectors $\underline{a}\to 0$ is trivially obtainable. In Section III, we follow the method initiated by Sharma to obtain the expansion coefficients for f(R) as an illustration of the general closed form for the expansion coefficients for $f(R)\ Y_L^M(\Theta,\Phi)$ recently obtained by Rashid[3]. The derivation of the results for this general case (which has already appeared) is omitted. In Section IV, we show how this generalization leads to the results obtained by Silverstone and Moats[2]. We emphasize the simplicity of Rashid's result for the special case when the displacement is along the z axis. In Section V we consider the limit when the displacement goes to zero. Though we are able to derive the desired limit, we find that the derivation is definitely non-trivial in comparison to what we see in the special case discussed in Section II. This perhaps calls for further and urgent progress on this problem.

C. A. Weatherford and H. W. Jones (eds.), ETO Multicenter Molecular Integrals, 61–72.

2. THE EXPANSION OF $R^L Y_L^M(\Theta,\Phi)$ ABOUT A DISPLACED CENTRE

In this section, we use the following notations:

$$\underline{r} = \underline{a} + \underline{R} \tag{1}$$

where ($\underline{r}$, $\underline{a}$, $\underline{R}$) have the spherical polar co-ordinates (r,θ,ϕ), (a,θ_a,ϕ_a) and (R,Θ,Φ), respectively. Then

$$\begin{aligned}\underline{r} &= r(\sin\theta\cos\phi,\ \sin\theta\sin\phi,\ \cos\theta)\\ \underline{a} &= a(\sin\theta_a\cos\phi_a,\ \sin\theta_a\sin\phi_a,\ \cos\theta_a)\\ \underline{R} &= R(\sin\Theta\cos\Phi,\ \sin\Theta\sin\Phi,\ \cos\Theta)\end{aligned} \tag{2}$$

which gives

$$R\sin\Theta\ e^{\pm i\Phi} = r\sin\theta e^{\pm i\phi} - a\sin\theta_a e^{\pm i\phi_a} \tag{3i}$$

and

$$R^2 = r^2+a^2 - 2ar(\cos\theta\cos\theta_a + \sin\theta\sin\theta_a\ \cos(\phi-\phi_a)) \tag{3ii}$$

To establish the expression for $R^L Y_L^M(\Theta,\Phi)$, we utilize[5]

$$\begin{aligned}Y_\ell^m(\theta,\phi) &= (-1)^m Y_\ell^{-m*}(\theta,\phi) = (-1)^m \sqrt{\frac{(2\ell+1)(\ell-m)!}{4\pi(\ell+m)!}}\ P_\ell^m(\cos\theta)\ e^{im\phi}\\ &= (-1)^m \sqrt{\frac{(2\ell+1)(\ell-m)!}{4\pi(\ell+m)!}}\ \frac{(1-x^2)^{m/2}}{2^\ell\cdot\ell!}\ \frac{d^{\ell+m}}{dx^{\ell+m}}\ (x^2-1)^\ell\ e^{im\phi}\end{aligned} \tag{4i}$$

which in the special case $m = \ell$ becomes

$$Y_\ell^\ell(\Theta,\Phi) = (-1)^\ell\ Y_\ell^{-\ell*}(\theta,\phi) = (-1)^\ell \sqrt{\frac{(2\ell+1)\,!}{4\pi}}\ \frac{1}{2^\ell\cdot\ell!}\ (\sin\Theta e^{i\phi})^\ell. \tag{4ii}$$

Thus

$$\begin{aligned}R^L Y_L^L(\Theta,\Phi) &= (-1)^L \sqrt{\frac{(2L+1)!}{4\pi}}\ \frac{1}{2^L\cdot L!}\ (R\sin\Theta\ e^{i\Phi})^L\\ &= (-1)^L \sqrt{\frac{(2L+1)\,!}{4\pi}}\ \frac{1}{2^L\cdot L!}\ (r\sin\theta\ e^{i\phi} - a\sin\theta_a\ e^{i\phi_a})^L\end{aligned}$$

$$= (-1)^L \sqrt{\frac{(2L+1)!}{4\pi}} \frac{1}{2^L \cdot L!} \sum_{\ell} (-1)^{\ell-L} \frac{L!}{\ell!(L-\ell)!} (r\sin\theta\, e^{i\phi})^{\ell}$$

$$\times\ (a\sin\theta_a\, e^{i\phi_a})^{-\ell+L}$$

or

$$R^L Y_L^L(\Theta,\Phi) = (-1)^L \sum_{\ell} (-1)^{\ell} \sqrt{\frac{4\pi(2L+1)!}{(2\ell+1)!\,(-2\ell+2L+1)!}}\ r^{\ell}\, Y_{\ell}^{\ell}(\theta,\phi)\ a^{-\ell+L} \times$$

$$Y_{-\ell+L}^{-\ell+L}(\theta_a,\phi_a). \tag{5}$$

In deriving the above equations, we utilized Eqs.(3i) and (4ii) and the Binomial theorem. Note the general convention used in this paper of not mentioning the range of summations in any expression explicitly wherein these are fixed by the non-negavity of the arguments of the factorials present in the expression with <u>integral</u> arguments.

In Eq.(5) above, we have the expansion for $R^L Y_L^M(\Theta,\Phi)$.

To obtain the corresponding general expansion for $R^L Y_L^M(\Theta,\Phi)$, we need to operate on the above by

$$J_-^{L-M} = (J_-^{(1)} + J_-^{(2)})^{L-M} = \sum_m \frac{(L-M)!}{(\ell-m)!\,(L-\ell-M+m)!} (J_-^{(1)})^{\ell-m}$$

$$\times\ (J_-^{(2)})^{-\ell+L-M+m} \tag{6}$$

and use

$$J_-^{L-M}\, Y_L^L(\theta,\phi) = \sqrt{\frac{(2L)!\,(L-M)!}{(L+M)!}}\ Y_L^M(\theta,\phi) \tag{7}$$

to arrive at

$$R^L Y_L^M(\Theta,\Phi) = (-1)^L \sum_{\ell m} (-1)^{\ell} \sqrt{\frac{4\pi(2L+1)!}{(2\ell+1)!\,(-2\ell+2L+1)!}} \langle \ell,m;\ -\ell+L,$$

$$-m+M|L,M\rangle\ (r^{\ell} Y_{\ell}^{m}(\theta,\phi))\,(a^{-\ell+L}\, Y_{-\ell+L}^{-m+M}(\theta_a,\phi_a)). \tag{8}$$

In the above expansion for $R^L Y_L^M(\Theta,\Phi)$ we note the following:

i) If we had $\underline{R} = \underline{r} + \underline{a}$ in place of $\underline{R} = \underline{r} - \underline{a}$ as in Eq.(1), we would have obtained the same expression as in the above except for the omission of $(-1)^{L+\ell}$ since $\underline{a} \to -\underline{a}$ results in $\theta_a,\ \phi_a \to \pi - \theta_a,\ \pi + \phi_a$ and

$$Y_{L-}^{M-m}(\pi - \theta_a, \pi + \phi_a) = (-1)^{L-} Y_{L-}^{M-m}(\theta_a, \phi_a).$$

ii)To arrive at the limit when $a \to 0$, we need just to note that when $a \to 0$, $a^{L-\ell} \to \delta_{L-\ell,0}$ and Eq.(8) immediately reduces to

$$R^L Y_L^M(\Theta, \Phi) = r^L Y_L^M(\theta, \phi)$$

as expected.

This feature is unfortunately not present in the expansions which appear in the following sections.

3. EXPANSION OF A GENERAL ORBITAL ABOUT A DISPLACED CENTRE

We shall try to compute the α's appearing in the expansion[6]

$$\frac{1}{R} f_{NL}(R) \; Y_L^M(\Theta, \Phi) = \sum_{\ell=0}^{\infty} \frac{1}{r} \alpha_\ell(NLM|a,r) \; Y_\ell^M(\theta, \phi) \tag{9}$$

where we are restricting $\underline{a}$ to be along the z axis, i.e. $\theta_a = \phi_a = 0$. Then $\Phi = \phi$ and

$$R^2 = r^2 + a^2 - 2ar\cos\theta \tag{10i}$$

or

$$\sin\theta d\theta = \frac{R}{ar} dR \tag{10ii}$$

To illustrate the method, we shall first specialize to the case L=M=0. Then $Y_0^0(\Theta, \Phi) = 1/\sqrt{4\pi}$. Using orthonormality of the spherical harmonics, namely[7]

$$\int_0^{2\pi} \int_0^{\pi} Y_{\ell m}^*(\theta, \phi) \; Y_{\ell' m'}(\theta, \phi) \sin\theta d\theta d\phi = \delta_{\ell \ell'} \delta_{mm'} \tag{11}$$

in Eq.(9) results in

$$\alpha_\ell(N00|a,r) = \frac{r}{\sqrt{4\pi}} \int \frac{f_{NL}(R)}{R} \; Y_\ell^0(\theta, \phi) \sin\theta d\theta d\phi \tag{12}$$

Since the integrand is independent of ϕ, the ϕ integration results in a factor of 2π. The θ integration is replaced by an integration over R using Eq.(12). We also use [8]

$$Y_\ell^o(\theta,\phi) = \sqrt{\frac{(2\ell+1)}{4\pi}} \frac{1}{2^\ell \cdot \ell!} \frac{d^\ell}{dx^\ell}(x^2-1)^\ell$$

$$= \sqrt{\frac{(2\ell+1)}{4\pi}} \cdot \frac{1}{2^\ell} \sum_p (-1)^{\ell-p} \frac{(2p)!}{p!\,(\ell-p)!\,(2p-\ell)!} x^{2p-\ell}$$

where

$$x = \cos\theta = \frac{r^2 + a^2 - R^2}{2ar}$$

which gives on utilizing the Binomial expansion

$$Y_\ell^o(\theta,\phi) = \sqrt{\frac{2\ell+1}{4\pi}} \sum_{pst} (-1)^{\ell-p+s} 2^{-2p} \frac{(2p)!\,(R^2/a^2)^s (r/a)^{2t-2p+\ell}}{p!(\ell-p)!\,s!\,t!\,(2p-\ell-s-t)!} .$$

Replacing $\ell+t-p$ by v results in

$$Y_\ell^o(\theta,\phi) = \sqrt{\frac{2\ell+1}{4\pi}} \sum_{vst} (-1)^{s+t+v} \frac{2^{-2\ell-2t+2v} (2\ell+2t-2v)!\,(R^2/a^2)^s (r/a)^{-\ell+2v}}{s!\,t!\,(-t+v)!(\ell+t-v)!(\ell-s+t-2v)!}$$

From the duplication formula (13)

$$\Gamma(2Z) = \frac{1}{\sqrt{\pi}} 2^{2Z-1} \Gamma(Z)\,\Gamma(Z+\tfrac{1}{2})$$

we find

$$(2\ell+2t-2v)! = \frac{1}{\sqrt{\pi}} 2^{2\ell+2t-2v} (\ell+t-v)!\,(\ell+t-v-\tfrac{1}{2})! \qquad (14)$$

(we shall be using the factorials even when the argument is not necessarily a non-negative integer. Indeed $x! = \Gamma(x+1)$).

Now the t summation in Eq.(13) can be performed using [9]

$$\sum (-1)^t \frac{(a+t)!}{t!(b+t)!\,(c-t)!} = (-1)^c \frac{a!(a-b)!}{c!\,(b+c)!\,(a-b-c)!} \qquad (15)$$

to arrive at

$$Y_\ell^o(\theta,\phi) = \sqrt{\frac{(2\ell+1)}{2\pi}} \sum_{s v} (-1)^s \frac{(-\tfrac{1}{2}+s+v)!\,(-\tfrac{1}{2}+\ell-v)!}{s!\,v!\,(\ell-s-v)!\,(-\tfrac{1}{2}+s)!} (R^2/a^2)^s \left(\frac{r}{a}\right)^{-\ell+2v} \tag{16}$$

which finally gives

$$\alpha_\ell(NOO|a,r) = \frac{1}{a}\sqrt{\frac{2\ell+1}{2\pi}} \sum_{s v} \frac{(-\tfrac{1}{2}-s)!\,(-\tfrac{1}{2}+s+v)!\,(-\tfrac{1}{2}+\ell-v)!}{s!\,v!\,(\ell-s-v)!\,(-\tfrac{1}{2})!} \times$$

$$\times \left(\frac{r}{a}\right)^{2v-\ell} \int_{|a-r|}^{a+r} f_{NL}(R)\,(R/a)^{2s}\,dR \tag{17}$$

The generalization of the above result has been obtained by Rashid (3) utilizing essentially the above procedure though the derivation is, as expected, much more involved. The final answer is

$$\alpha_\ell(NLM|a,r) = \frac{1}{a} \sum_v (r/a)^{2v-\ell} \sum_s b_v(s\ell LM) \int_{|a-r|}^{a+r} f_{NL}(R)\,(R/a)^{2s-L} \tag{18}$$

where

$$b_v(s\ell LM) = \frac{(-1)^L}{2\pi}\sqrt{(2\ell+1)(2L+1)(\ell+M)!\,(\ell-M)!\,(L+M)!\,(L-M)!} \times$$

$$\times \frac{(-\tfrac{1}{2}+\ell-v)!\,(-\tfrac{1}{2}+L-s)!}{s!\,v!\,(\ell+L-s-v)!} \sum_t \frac{(-\tfrac{1}{2}-M+s-t+v)!}{t!\,(\ell-M-t)!\,(L-M-t)!\,(-\tfrac{1}{2}-M-t)!\,(2M+t)!} \tag{19}$$

The above expression for $b_v(s\ell LM)$ is a generalization of

$$b_v(s\ell oo) = \sqrt{\frac{2\ell+1}{2\pi}} \frac{(-\tfrac{1}{2}-s)!\,(-\tfrac{1}{2}+s+v)!\,(-\tfrac{1}{2}+\ell-v)!}{s!\,v!\,(\ell-s-v)!\,(-\tfrac{1}{2})!} \tag{20}$$

appearing in Eq.(17).

The b_v's appearing in Eq.(19) have the obvious symmetries

$$b_v(s\ell LM) = b_v(s\ell L(-M)) = b_v(sL\ell M)(-1)^{\ell+L} \tag{21}$$

Our expression for the b_v's contains only one summation. This may be compared with that given in Sharma's which contains 4 summations.

4. DERIVATION OF THE EXPANSION OBTAINED BY SILVERTSONE AND MOATS

Our notation which follows Sharma's work differs slightly from that used by Silverstone and Moats. In order to obtain the comparison, we have to replace our a and f(r) by R and rf(r). Then by taking θ_a, $\phi_a \to 0,0$ in Eq.(1) in Silverstone and Moats' paper[2], we find that

$$\alpha_\ell(NLM|a,r) = \sum_{\lambda=|-\ell+L|}^{\ell+L} V_{\ell\lambda L}(r,a)\, C_{\lambda LM\ell M}\, Y_\lambda^o(o,o) \tag{22}$$

where

$$Y_\lambda^o(o,o) = \sqrt{\frac{2\lambda+1}{4\pi}}$$

and

$$C_{\lambda LM\ell M} = \int d\Omega\, Y_\lambda^{o*}(\theta,\phi)\, Y_\ell^{M*}(\theta,\phi)\, Y_L^M(\theta,\phi)$$

$$= (-1)^M \sqrt{\frac{(2\ell+1)(2L+1)}{4\pi(2\lambda+1)}}\quad \langle\ell,M;L,-M|\lambda,o\rangle\langle\ell,o;L,o|\lambda,o\rangle \tag{23}$$

or using the orthogonality of the Clebsch-Gordon coefficients we find

$$\sqrt{\frac{(2\ell+1)(2L+1)}{4\pi}}\, V_{\ell\lambda L}(r,a)\, \langle\ell,o;L,o|\lambda,o\rangle$$

$$= \sum_M (-1)^M \langle\ell,M;L,-M|\lambda,o\rangle\, \alpha_\ell(NLM|a,r) \tag{24}$$

Since $V_{\ell\lambda L}(r,a)$ do not depend upon the angles (θ_a,ϕ_a), once these coefficients have been obtained from the α's by using the above equations, one can immediately write the general expansion given by Silverstone and Moats. In particular, for L=M=0, Eq.(24) becomes

$$\sqrt{\frac{2\ell+1}{4\pi}}\, V_{\ell\ell o}(r,a) = \alpha_\ell(NOO|a,r) \tag{25}$$

which is verified on noting that

$$D_{\ell\lambda L s t} = (-1)^{\frac{1}{2}(\ell+L+\lambda)}\, \frac{(-\frac{1}{2}+\ell-t)!\,(-\frac{1}{2}+L-s)!\,(-\frac{1}{2}-\frac{\ell+L-\lambda}{2}+s+t)!}{\pi^{3/2}\, s!\, t!\, (\frac{\ell+L+\lambda}{2}-s-t)!} \tag{26}$$

and

$$V_{\ell\ell o}(r,a) = 2\pi(-1)^{\ell} \frac{1}{a} \sum_{\mathcal{s}t} D_{\ell\ell o\mathcal{s}t} \left(\frac{r}{a}\right)^{2t-\ell} \int_{|r-a|}^{r+a} dr' \left(\frac{r'}{a}\right)^{2\mathcal{s}-L} f(r').$$

In the above, the expression for $D_{\ell\lambda L\mathcal{s}t}$ is equivalent to Eq.(4) and $V_{\ell\ell o}(r,a)$ is obtained from Eq.(3) in Ref.2. Note that when L=M=0, $\lambda=\ell$.

In order to obtain $V_{\ell\lambda L}$ from Eq.(24), we first compute

$$A = \sum_M (-1)^M b_v(\mathcal{s}\ell LM) \langle \ell,M;L,-M|\lambda,o\rangle$$

$$= \sum_M (-1)^M b_v(\mathcal{s}\ell LM) \langle L,M;\ell,-M|\lambda,o\rangle$$

$$= \frac{(-1)^L}{2\pi} \sqrt{(2\ell+1)(2L+1)} \frac{(-\frac{1}{2}+\ell-v)!\,(-\frac{1}{2}+L-s)!}{\mathcal{s}!\,v!\,(\ell+L-\mathcal{s}-v)!} \times$$

$$\times \sum_{tM} \frac{(-\frac{1}{2}-L+s+v+t)!}{t!\,(\ell-L+t)!\,(L-M-t)!\,(-\frac{1}{2}-L+t)!\,(L+M-t)!} \times$$

$$\times \sqrt{\frac{(\ell+L-\lambda)!\,(\ell-L+\lambda)!\,(-\ell+L+\lambda)!\,(2\lambda+1)\lambda!\,\lambda!}{(\ell+L+\lambda+1)!}}$$

$$\times \sum_Z (-1)^{\ell+Z} \frac{(\ell+\lambda+M-Z)!\,(L-M+Z)!}{Z!\,(\ell-L+\lambda-Z)!\,(\lambda-Z)!\,(-\ell+L+Z)!}$$

Now the M summation is given by [10]

$$\sum_M \frac{(\ell+\lambda+M-Z)!\,(L-M+Z)!}{(L+M-t)!\,(L-M-t)!} = \frac{(\ell+\lambda+L+1)!\,(\ell-L+\lambda-Z+t)!\,(Z+t)!}{(2L-2t)!\,(\ell-L+\lambda+1+2t)!}$$

which shows that

$$A = (-1)^{\ell+L} \frac{1}{2\pi} \sqrt{(2\ell+1)(2L+1)} \frac{(-\frac{1}{2}+\ell-v)!\,(-\frac{1}{2}+L-\mathcal{s})!}{\mathcal{s}!\;v!\;(\ell+L-\mathcal{s}-v)!} \times$$

$$\times \sqrt{(2\lambda+1)\,(\ell+L-\lambda)!\,(\ell-L+\lambda)!\,(-\ell+L+\lambda)!\,(\ell+L+\lambda+1)!} \quad \times$$

$$\times \sum_{tz}(-1)^{z} \frac{(t+z)!\,(\ell-L+\lambda+t-Z)!\,(-\tfrac{1}{2}+L+s+v+t)!}{t!\,(\ell-L+t)!\,(-\tfrac{1}{2}-L+t)!\,(2L-2t)!\,(\ell-L+\lambda+1+2t)\,!} \quad \times$$

$$\times \frac{1}{Z!\,(\ell-L+\lambda-Z)!\,(\lambda-Z)!\,(-\ell+L+Z)\,!} \quad .$$

Next we perform the z summation using a modified form of Dixon's theorem[11]

$$\sum_{z}(-1)^{Z} \frac{(t+Z)!\,(\ell-L+\lambda+t-Z)!}{Z!\,(\ell+\lambda-L-Z)!\,(\lambda-Z)!\,(-\ell+L+Z)!}$$

$$= 2\,\frac{t!\quad(\ell-L+t)!\,(-\ell+L+\lambda-1)!\qquad(\tfrac{1}{2}(\ell-L+\lambda)+t)!}{\lambda!\quad(-\ell+L+\lambda)!\,\left(\frac{-\ell+\lambda+L}{2}-1\right)!\,\left(\frac{\ell-L+\lambda}{2}\right)!\,(\tfrac{1}{2}(\ell-L-\lambda)+t)!} \quad .$$

Next the t summation can be done using[12]

$$\sum(-1)^{t} \frac{(a-t)!}{t!\,(b-t)!\,(c-t)!} = \frac{(a-b)!\,(a-c)!}{b\,!\,c\,!\,(a-b-c)!}$$

which finally gives on repeated use of the duplication formula for the gamma function and the relation

$$\Gamma(Z)\Gamma(1-Z) = \pi/\sin\pi Z$$

the expression for $V_{\ell\lambda L}(r,a)$ as in Ref.2. Thus we have established the equivalence of the two expressions.

We wish however to note that the expression in Silverstone and Moats which can be obtained from ours contains the expansion for the case when the displacement vector $\underline{a}$ is not necessarily along the z axis. But if we wish to use it for the case where $\underline{a}$ is along the z axis, we are left with two summations, namely the one over λ and the one implicit $C_{\lambda LM\ell M}$, besides the two over s and t in the $D_{\ell\lambda Lst}$, whereas our corresponding expression for α contains the two over s and v but just one additional one over t. Thus, in this case, our expansion is much easier to handle and may prove to be more advantageous in numerical calculations for various types of orbitals.

5. THE LIMIT $a \to 0$ OF $\alpha_\ell(NLM|a,r)$

When $a \to 0$, the displacement disappears. Thus $\underline{R} = \underline{r}$ and the expression in Eq.(9) must reduce to an identity, i.e.

$$\underset{a\to 0}{\ell t}\ \alpha_{\ell}(NLM\,a,r) = \delta_{\ell L} \tag{27}$$

independently of M. However, though as remarked at the end of Section II, the expansion in Eq.(8) for $R^L Y_L^M(\Theta,\Phi)$ does satisfy this condition trivially, to obtain the limit $a \to 0$ for the coefficients $\alpha_{\ell}(NLM|a,r)$ given in Eqs.(18) and (19) does not seem to be so trivial. Even for the special case of $\alpha_{\ell}(N00|a,r)$ given in Eq.(17), the limiting procedure is not at all manifest. The reason why this problem is arising is in the transformation from the angle θ to the radius vector R given in Eq.(10i) which becomes singular when $a\to 0$. In the following we shall demonstrate how this limit can be retrieved though the calculations unfortunately are lengthy and involved.

Let us write

$$F(R) = \int f(R)\ (R/a)^{2s-L}\ dR \tag{28}$$

where F(R) is an <u>indefinite</u> integral of $f(R)\left(\frac{R}{a}\right)^{2s-L}$. Then

$$\int_{r-a}^{r+a} f(R)\ (R/a)^{2s-L}\ dR = F(r+a)-F(r-a) \tag{29}$$

(Note that we wish to consider $a\to 0$. The lower limit $|r-a|$ has therefore been replaced by r-a).

We expand the right-hand side in the above equation by using Taylor's theorem to arrive at

$$\int_{r-a}^{r+a} f(R)\ (R/a)^{2s-L}\ dR$$

$$= 2a \sum_{p=0} \sum_{k=[p/2]} \frac{(2s-L)!}{(2k+1)!\,(-L+2s-2k+p)!}\ {}^{2k}C_p\,(r/a)^{-L+2s-2k+p}\ \left(a\frac{d}{dr}\right)^p f(r) \tag{30}$$

Substituting the above in $\alpha_{\ell}(NLM|a,r)$ we find

$$\alpha_{\ell}(NLM|a,r)$$

$$= 2\sum_{svpk} \frac{b_v(s\ell LM)\,(2s-L)!}{(2k+1)!\,(-L+2s-2k+p)!}\ {}^{2k}C_p\left(\frac{r}{a}\right)^{-\ell-L+2s+2v-2k+p}\left(a\frac{d}{dr}\right)^p f(r)$$

The s-v summation occurs in the above in

$$A_{kp} = \sum_{sv} \frac{(-\tfrac{1}{2}+\ell-v)!\,(-\tfrac{1}{2}+L-s)!\,(-\tfrac{1}{2}-M+s+v-t)!\,(2s-L)!}{s!\ v!\ \ (\ell+L-s-v)!\,(-L+2s-2k+p)!\,(2k+1)!} \times$$

$$\times\ (r/a)^{-\ell-L+2s+2v-2k+p}$$

In the above let us replace $\ell+L-s-v$ by S. Then

$$A_{kp} = \sum_{sS} \frac{(-\tfrac{1}{2}+L-S)!\ (-\tfrac{1}{2}-L+s+S)!\ (-\tfrac{1}{2}-M+\ell+L-S-t)!\ (2s-L)!}{S!\ s!\ \ (\ell+L-s-S)!\ (2k+1)!\ (2s-L-2k+p)!}$$

$$\times\ \left(\frac{r}{a}\right)^{\ell+L-2k-2S+p}$$

The limit exists when $\ell+L-2k-2S+p = 0$ and is indeed finite when $a \to 0$ and the factor $(a\frac{d}{dr})^p f(r)$ in $\alpha_\ell(NLM|a,r)$ is zero except for p=0.

Thus we shall be left with p=0 case only. The summation over s can be performed using the duplication formula and Minton-Karlsson theorem[13,14] to give

$$A_{ko} = (-1)^{\frac{1}{2}(\ell-L)}\ \pi\ \frac{(-\tfrac{1}{2}-M+\frac{\ell+L}{2}+k-t)!\ \ 2^{2k}}{(2k+1)!\ (\frac{\ell+L}{2}-k)!}$$

from which we find using Eq.(15)

$$\sum_{k=o}^{\frac{1}{2}(\ell+L)} A_{ko} = (-1)^{\frac{1}{2}(\ell-L)}\ \frac{\pi^{3/2}}{2}\ \sum_k (-1)^k\ \frac{(-\tfrac{1}{2}-M+\frac{\ell+L}{2}+k-t)!}{k!\ (\tfrac{1}{2}+k)!\ (\frac{\ell+L}{2}-k)!}$$

$$= (-1)^{\frac{1}{2}(\ell-L)}\ \frac{\pi^{3/2}}{2}\ \frac{(-\tfrac{1}{2}-M+\frac{\ell+L}{2}-t)!\ (M+t)!}{(\frac{\ell+L}{2})!\ (\frac{\ell+L+1}{2})!\ (-\frac{[\ell+L]}{2}+M+t)!}\ .$$

Finally we have to perform the t summation after dividing by

$$\left[t!\ (\ell-M-t)!\ (L-M-t)!\ (-\tfrac{1}{2}-M-t)!\ (2M+t)!\right]$$

Noting the factorials $\ell-M-t$, $L-M-t$, $-\frac{\ell+L}{2}+M+t$ we see that

$$(\ell-M-t) + (L-M-t) + 2(-\frac{[\ell+L]}{2}+M+t) = 0$$

wherein each of the bracketed quantities is non-negative. Thus we deduce that $\ell=L$, $t=\ell-M$. Collecting the various factors in $\alpha_\ell(NLM|a,r)$ we finally arrive at

$$\underset{a \to o}{\delta t}\ \alpha_{\ell}(NLM|a,r) = \delta_{\ell L}$$

as required.

6. ACKNOWLEDGEMENTS

The author would like to thank Professor Abdus Salam, the International Atomic Energy Agency and UNESCO for hospitality at the International Centre for Theoretical Physics, Trieste.

REFERENCES

1. Sharma, R.R.: 1976, Phys. Rev. A 13, pp. 517-527.

2. Silverstone, H.J. and Moats, R.K.: 1977, Phys. Rev. A 16, pp. 1723-32.

3. Rashid, M.A.: 1981, J. Math. Phys. 22, pp.271-74.

4. Brink, D.M. and Satchler, G.R.: 1968, Angular Momentum, Oxford University Press, Oxford.

5. Edmonds, A.R.: 1960, Angular Momentum in Quantum Mechanics, Princeton University Press, Princeton, N.J., Eqs. (2.5.6 and 29).

6. We are using the notations of Ref. 1.

7. Eq. (2.5.4), p.21 in Ref. 5.

8. This is a special case of Eq.(4i) of this paper.

9. Eq. (A1.2), p.121 in Ref. 5.

10. Eq. (A1.3), p.121 in Ref. 5.

11. Bailey, W.N.: 1964, Generalized Hypergeometric Series, Stechert-Harner Service Agency, New York, Eq. (3.1(1)), p.13.

12. This is a modified form of Eq. (A1.2) in Ref. 5.

13. Minton, B.M.: 1970, J. Math. Phys. 11, p.1375.

14. Karlsson, P.W.: 1971, J. Math. Phys. 12, p.270.

RECENT PROGRESS ON THREE-CENTER INTEGRALS USING FOURIER-TRANSFORM-BASED ANALYTICAL FORMULAS

H. Davis Todd
Department of Chemistry
Wesleyan University, Middletown, CT 06457

Kenneth C. Daiker
CONDUIT, Box 388
Iowa City, Iowa 52244

Robert D. Cloney, S.J.
Department of Chemistry
Fordham University
Bronx, New York 10458

Richard K. Moats[a)] and Harris J. Silverstone
Department of Chemistry
The Johns Hopkins University
Baltimore, Maryland 21218

Recent progress on computer evaluation of three-center integrals of $1/r_{12}$ with Slater-type orbitals is described. Fourier-transform-based formulas are first put in a programmable form. Then the special functions that occur are evaluated by either explicit expressions or recursion formulas. Convergence rate and asymptotics are touched upon briefly.

I. INTRODUCTION

Over thirteen years ago the formal advantages of using Fourier transforms for evaluating $1/r_{12}$ integrals with exponential-type orbitals became apparent. All three- and four-center integrals could be represented as infinite sums of terms involving spherical harmonics, spherical Bessel functions, and exponential-type integrals.[1-9] The resulting formulas, while simple in an analytic sense, were sufficiently complicated in a practical sense that only the two-center-integral formulas[8] and the asymptotic formulas[7] for the multicenter integrals have been incorporated into "production" computer programs.[10,11] In the last few years we have resumed the development of programs for Fourier-transform-based formulas for multicenter integrals, and this article is to summarize the current status of that effort, particularly with regard to the (1-2)-type three-center integral.[12]

C. A. Weatherford and H. W. Jones (eds.), ETO Multicenter Molecular Integrals, 73–89.

In Sec. II the Fourier-transform application is briefly reviewed, with emphasis on the (1-2)-type three-center two-electron integral. In Secs. III and IV the derivative operators that appear in the formulas are eliminated in favor of summations and special functions. Computer programs are discussed in Sec. V, convergence and asymptotics in Sec. IV.

II. BRIEF REVIEW OF FOURIER-TRANSFORM-BASED FORMULAS

The Fourier-transform convolution theorem makes a six-dimensional coordinate-space Coulomb-interaction integral of two charge distributions into a three-dimensional integral over the Fourier transforms:

$$\int dV_1 \int dV_2 \; \rho_1(\vec{r}_1)^* \rho_2(\vec{r}_2-\vec{R}) \; 1/r_{12}$$

$$= (2\pi)^{-3} \int d^3k \; \{F.T.\rho_1\}^* \{F.T.\rho_2\} \{F.T.r^{-1}\} e^{i\vec{k}\cdot\vec{R}} \quad . \tag{1}$$

Formulas for evaluating the Fourier transforms of $1/r$, of a Slater-type orbital (STO), and of a two-center product of STO's are known, and the end result is a formula[1-9] for the multicenter integral that is like a multipole expansion:

$$I(\vec{R}) = \{ \sum_{[\ell]} \sum_{[m]} \} \{ \langle Y_{\ell_1}^{m_1} Y_{\ell_2}^{m_2} Y_{\ell_3}^{m_3} \rangle \} \{ Y_\lambda^m(\theta_R, \phi_R) \} \; I^{[\ell]}(R) \quad . \tag{2}$$

The four elements of this formula are (i) a sum over ℓ- and m-type quantum numbers, (ii) factors that are the integral of triple products of spherical harmonics, (iii) factors that are spherical harmonics of the angular coordinates of internuclear-distance vectors, and (iv) "radial" factors that depend on internuclear distances of ℓ-type quantum numbers (but not on m-type quantum numbers or on angles). The difficult part is to evaluate the radial factors $I^{[\ell]}$

The radial factors yield to contour-integration techniques of complex variable theory, and the resulting formulas can be expressed in terms of four "ordinary" special functions, [2,13] two that are of the modified spherical Bessel function type, and two that are of the exponential-integral type:

$$K_\ell(x) = (-x)^\ell (x^{-1} d/dx)^\ell x^{-1} e^{-x} \quad , \tag{3}$$

$$I_\ell(x) = x^\ell (x^{-1} d/dx)^\ell x^{-1} \sinh x \quad , \tag{4}$$

$$E_n(x) = \int_1^\infty dt\ t^{-n} e^{-xt} = \alpha_{-n}(x) \quad , \tag{5}$$

$$\tilde{E}_n(x) = E_n(x)+(-x)^{n-1}[\log x-\psi(n)]/(n-1)! \quad , (n-1\geq 0) \quad , \tag{6}$$

$$= \alpha_{-n}(x)-x^{n-1}(-n)! = \hat{\alpha}_{-n}(x) \quad , (-n\geq 0) \quad . \tag{7}$$

In this article we focus on the (1-2)-type three-center integral of $1/r_{12}$, which is explicitly characterized by the formulas[2]

$$\int dV_1 \int dV_2\ \psi_c^*(\vec{r}_1)\psi_a^*(\vec{r}_2-\vec{R})\psi_b(\vec{r}_2-\vec{R}-\vec{\mathcal{R}})\ r_{12}^{-1}$$

$$= (-1)^{\ell_c}\pi^3\{\sum_\ell \sum_{\Lambda=|\ell-\ell_a|}^{\ell+\ell_a} \sum_{\lambda=|\ell-\ell_b|}^{\ell+\ell_b} \sum_{t=|\Lambda-\ell_c|}^{\Lambda+\ell_c} \sum_m\}$$

$$\times \langle Y_\ell^{m*} Y_{\ell_a}^{m_a} Y_\Lambda^{m-m_a} \rangle \langle Y_{\ell_b}^{m_b *} Y_\ell^m Y_\lambda^{m_b-m} \rangle \langle Y_\Lambda^{m-m_a *} Y_{\ell_c}^{m_c} Y_t^{m-m_a-m_c} \rangle$$

$$\times Y_\lambda^{m_b-m}(\theta_{\mathcal{R}},\phi_{\mathcal{R}}) Y_t^{m-m_a-m_c}(\theta_R,\phi_R)$$

$$\times I_{c;ab}^{t\ell\lambda\Lambda}(R,\mathcal{R}) \quad , \tag{8}$$

$$I_{c;ab}^{t\ell\lambda\Lambda}(R,\mathcal{R})/[8(-1)^\Lambda \mathcal{R}^{n_a+n_b} R^{n_c+2} (\mathcal{R}/R)^{\ell-n_a}]$$

$$= -\ [\cdots(-1)^\lambda I_\lambda(\zeta_b\mathcal{R})\cdots]_1[\cdots(-1)^t I_t(\zeta_c R)\cdots]_2$$

$$\times E_{\Lambda+\ell+1-n_a}[(\zeta_a+\zeta_b+\zeta_c)R]+[\cdots(-1)^\lambda I_\lambda(\zeta_b\mathcal{R})\cdots]_1[\cdots K_\lambda(\zeta_c R)\cdots]_2$$

$$\times \{(R/\mathcal{R})^{\Lambda+\ell-n_a}\tfrac{1}{2}\tilde{E}^{(2)}_{\Lambda+\ell+1-n_a}[(\zeta_a+\zeta_b\pm\zeta_c)\mathcal{R}]-\tfrac{1}{2}\tilde{E}^{(2)}_{\Lambda+\ell+1-n_a}[(\zeta_a+\zeta_b\pm\zeta_c)R]\}$$

$$+ [\cdots K_\lambda(\zeta_b \mathcal{R})\cdots]_1 [\cdots K_t(\zeta_c R)\cdots]_2$$

$$\times (R/\mathcal{R})^{\Lambda+\ell-n_a} \tfrac{1}{4}\tilde{E}^{(4)}_{\Lambda+\ell+1-n_a}[(\zeta_a \pm \zeta_b \pm \zeta_c)\mathcal{R}]$$

$$+\delta_{\Lambda,t+\ell_c} \frac{(n_c+\ell_c+1)!(2\Lambda-1)!!}{(2\ell_c+1)!!(2t+1)!!} (\zeta_c R)^{-n_c-\ell_c-2}$$

$$\times [\cdots(-1)^\lambda I_\lambda(\zeta_b \mathcal{R})\cdots]_1 \, E_{\Lambda+\ell+1-n_a}[(\zeta_a+\zeta_b)R]$$

$$+\delta_{t,\Lambda+\ell_c} (-1)^{\ell_c} \frac{(n_c+\ell_c+1)!(2t-1)!!}{(2\ell_c+1)!!(2\Lambda+1)!!} (\zeta_c R)^{-n_c-\ell_c-2}$$

$$\times([\cdots(-1)^\lambda I_\lambda(\zeta_b \mathcal{R})\cdots]_1$$

$$\times \{(\mathcal{R}/R)^{n_a+\Lambda-\ell+1} \alpha_{n_a+\Lambda-\ell}[(\zeta_a+\zeta_b)\mathcal{R}] - \alpha_{n_a+\Lambda-\ell}[(\zeta_a+\zeta_b)R]\}$$

$$+[\cdots K_\lambda(\zeta_b \mathcal{R})\cdots]_1 \, (\mathcal{R}/R)^{n_a+\Lambda-\ell+1} \tfrac{1}{2}\hat{\alpha}^{(2)}_{n_a+\Lambda-\ell}[(\zeta_a \pm \zeta_b)\mathcal{R}]) \;, (R \geq \mathcal{R}), \quad (9)$$

$$= -[\cdots(-1)^\lambda I_\lambda(\zeta_b \mathcal{R})\cdots]_1 [\cdots(-1)^t I_t(\zeta_c R)\cdots]_2$$

$$\times (R/\mathcal{R})^{\Lambda+\ell-n_a} E_{\Lambda+\ell+1-n_a}[(\zeta_a+\zeta_b+\zeta_c)\mathcal{R}]$$

$$+[\cdots K_\lambda(\zeta_b \mathcal{R})\cdots]_1 [\cdots(-1)^t I_t(\zeta_c R)\cdots]_2$$

$$\times\{\tfrac{1}{2}\tilde{E}^{(2)}_{\Lambda+\ell+1-n_a}[(\zeta_a \pm \zeta_b+\zeta_c)R] - (R/\mathcal{R})^{\Lambda+\ell-n_a} \tfrac{1}{2}\tilde{E}^{(2)}_{\Lambda+\ell+1-n_a}[(\zeta_a \pm \zeta_b+\zeta_c)\mathcal{R}]\}$$

$$+ [\cdots K_\lambda(\zeta_b \mathcal{R})\cdots]_1 [\cdots K_t(\zeta_c R)\cdots]_2 \tfrac{1}{4}\tilde{E}^{(4)}_{\Lambda+\ell+1-n_a}[(\zeta_a \pm \zeta_b \pm \zeta_c)R]$$

$$+\delta_{\Lambda,t+\ell_c} \frac{(n_c+\ell_c+1)!(2\Lambda-1)!!}{(2\ell_c+1)!!(2t+1)!!} (\zeta_c R)^{-n_c-\ell_c-2}$$

$$\times \; [\cdots(-1)^{\lambda} I_{\lambda}(\zeta_b \mathcal{R})\cdots]_1 \; (R/\mathcal{R})^{\Lambda+\ell-n_a} E_{\Lambda+\ell+1-n_a}[(\zeta_a+\zeta_b)\mathcal{R}]$$

$$+[\cdots K_{\lambda}(\zeta_b \mathcal{R})\cdots]_1$$

$$\times \; (R/\mathcal{R})^{\Lambda+\ell-n_a} \tfrac{1}{2}\tilde{E}^{(2)}_{\Lambda+\ell+1-n_a}[(\zeta_a\pm\zeta_b)\mathcal{R}] - \tfrac{1}{2}\tilde{E}^{(2)}_{\Lambda+\ell+1-n_a}[(\zeta_a\pm\zeta_b)R]\}$$

$$+\delta_{t,\Lambda+\ell_c} (-1)^{\ell_c} \frac{(n_c+\ell_c+1)!(2t-1)!!}{(2\ell_c+1)!!(2\Lambda+1)!!} (\zeta_c R)^{-n_c-\ell_c-2}$$

$$\times[\cdots K_{\lambda}(\zeta_b \mathcal{R})\cdots]_1 \; \tfrac{1}{2}\hat{\alpha}^{(2)}_{n_a+\Lambda-\ell}[(\zeta_a\pm\zeta_b)R] \quad , \quad (R\leq\mathcal{R}) \quad , \qquad (10)$$

$$[\cdots F_{\lambda}(x)\cdots]_1 = (-d/dx)^{n_b-\ell_b}(x^{-1}d/dx)^{\ell_b} x^{\ell+\ell_b+1} F_{\lambda}(x)(x^{-1}d/dx)^{\ell} x^{-1} , \qquad (11)$$

$$[\cdots F_t(x)\cdots]_2 = (-d/dx)^{n_c-\ell_c}(x^{-1}d/dx)^{\ell_c} x^{\ell_c+\Lambda-1} F_t(x)(x^{-1}d/dx)^{\Lambda} x^{-1} , \qquad (12)$$

$$\tilde{E}^{(2)}_n(x\pm y) = \tilde{E}^{(2)}_n(x+Y)-\tilde{E}^{(2)}_n(x-y) = \hat{\alpha}^{(2)}_{-n}(x\pm y) \quad , \qquad (13)$$

$$\tilde{E}^{(4)}_n(x\pm y\pm z) = \tilde{E}_n(x+y+z)-\tilde{E}_n(x-y+z)-\tilde{E}_n(x+y-z)+\tilde{E}_n(x-y-z) \quad . \qquad (14)$$

Much of the discussion here pertains to the (1-2)-type integral with a more general $r_{12}^N Y_L^M(\theta_{12},\phi_{12})$ interaction (vs. $1/r_{12}$) , as well as to other types of multicenter integrals. Note that the above formulas are completely elementary from the viewpoint of analysis, but not from the viewpoint of computation, because of the various derivatives. The route to practical formulas is sketched in the next section.

III. ELIMINATION OF DERIVATIVES IN FORMULAS FOR THE (1-2)-TYPE THREE-CENTER INTEGRAL

Before Eqs. (8)-(14) can be programmed for a computer, the derivatives must be explicitly taken. Our approach has been developed in a previous paper[8] on two-center integrals. The key ideas are (i) that the derivative of a spherical Bessel function is a sum of spherical Bessel functions, and (ii) that it makes sense to define new "special functions" by the action of operators of the form $(x^{-1}d/dx)^{\ell}x^{-1}$ acting on exponential-type integrals. In the case of the two-center integrals, only two "two-parameter" special functions occurred: $A^{\ell}_{n}(x,y)$ and $\hat{A}^{\ell}_{n}(x,y)$. In the present case, these same two functions occur in both their original form and in a version in which the subscript is negative. In addition, certain analogous "three-parameter" special functions arise. These two- and three-parameter special functions are

$$A^{\ell}_{n}(x,y) = (-x)^{\ell}(x^{-1}d/dx)^{\ell}x^{-1}\alpha_{n}(x+y) = E^{\ell}_{-n}(x,y) \quad , \tag{15}$$

$$\hat{A}^{\ell}_{n}(x,y) = x^{\ell}(x^{-1}d/dx)^{\ell}x^{-1}\tfrac{1}{2}\hat{\alpha}^{(2)}_{n}(y\pm x) = \tilde{E}^{\ell}_{-n}(x,y) \quad , \tag{16}$$

$$E^{\ell_1\ell_2}_{n}(x,y,z) = (-x)^{\ell_1}(x^{-1}d/dx)^{\ell_1}x^{-1}(-y)^{\ell_2}(y^{-1}d/dy)^{\ell_2}y^{-1}E_{n}(x+y+z), \tag{17}$$

$$\tilde{E}^{\ell_1\ell_2}_{n}(x,y,z) = (-x)^{\ell_1}(x^{-1}d/dx)^{\ell_1}x^{-1}y^{\ell_2}(y^{-1}d/dy)^{\ell_2}y^{-1}\tfrac{1}{2}\tilde{E}^{(2)}_{n}(z+x\pm y), \tag{18}$$

$$\tilde{E}^{\ell_1\ell_2}_{n}(x,y,z) = (-x)^{\ell_1}(x^{-1}d/dx)^{\ell_1}x^{-1}y^{\ell_2}(y^{-1}d/dy)^{\ell_2}y^{-1}\tfrac{1}{4}\tilde{E}^{(4)}_{n}(z\pm x\pm y), \tag{19}$$

With the aid of the two- and three-parameter special functions, whose evaluation will be discussed in the next section, the radial factors can be put in a readily programmable form:

$$I^{t\ell\lambda\Lambda}_{c;ab}(R,\mathcal{R})/[8(-1)^{\Lambda}\mathcal{R}^{n_a+n_b}R^{n_c+2}(\mathcal{R}/R)^{\ell-n_a}]$$

$$=S^{\lambda\ell}_{b1}(\zeta_b\mathcal{R})\,S^{t\Lambda}_{c2}(\zeta_c R)$$

$$\times\Big\{-(-1)^{u_b+u_c+\ell_b+\ell_c}\,I_{\lambda+t_b}(\zeta_b\mathcal{R})\,I_{t+t_c}(\zeta_c R)$$

$$\times (R/\mathcal{R})^{\ell+u_b+1} E^{\ell+u_b,\Lambda+u_c}_{\Lambda+\ell+1-n_a}(\zeta_b R,\zeta_c R,\zeta_a R)$$

$$+(-1)^{u_b+\ell_b+t_c} \mathcal{I}_{\lambda+t_b}(\zeta_b \mathcal{R}) \mathcal{K}_{t+t_c}(\zeta_c R)$$

$$\times [(R/\mathcal{R})^{\ell-n_a-u_c-1} \tilde{E}^{\ell+u_b,\Lambda+u_c}_{\Lambda+\ell+1-n_a}(\zeta_b \mathcal{R},\zeta_c \mathcal{R},\zeta_a \mathcal{R})$$

$$-(R/\mathcal{R})^{\ell+u_b+1} \tilde{E}^{\ell+u_b,\Lambda+u_c}_{\Lambda+\ell+1-n_a}(\zeta_b R,\zeta_c R,\zeta_a R)]$$

$$+(-1)^{t_b+t_c} \mathcal{K}_{\lambda+t_b}(\zeta_b \mathcal{R}) \mathcal{K}_{t+t_c}(\zeta_c R)$$

$$\times (R/\mathcal{R})^{\ell-n_a-u_c-1} \tilde{\mathbb{E}}^{\ell+u_b,\Lambda+u_c}_{\Lambda+\ell+1-n_a}(\zeta_b \mathcal{R},\zeta_c \mathcal{R},\zeta_a \mathcal{R}) \Big\}$$

$$+ \mathcal{S}^{\lambda\ell}_{b1}(\zeta_b \mathcal{R}) \frac{(n_c+\ell_c+1)!}{(2\ell_c+1)!!} (\zeta_c R)^{-n_c-\ell_c-2}$$

$$\times \Big\{ \delta_{\Lambda,t+\ell_c} \frac{(2\Lambda-1)!!}{(2t+1)!!} (-1)^{\ell_b+u_b} \mathcal{I}_{\lambda+t_b}(\zeta_b \mathcal{R})$$

$$\times (R/\mathcal{R})^{\ell+u_b} E^{\ell+u_b}_{\Lambda+\ell+1-n_a}(\zeta_b R,\zeta_a R)$$

$$+\delta_{t,\Lambda+\ell_c} \frac{(2t-1)!!}{(2\Lambda+1)!!} \Big((-1)^{\ell_c+\ell_b+u_b} \mathcal{I}_{\lambda+t_b}(\zeta_b \mathcal{R})$$

$$\times [(R/\mathcal{R})^{\ell-n_a-\Lambda-1} A^{\ell+u_b}_{n_a+\Lambda-\ell}(\zeta_b \mathcal{R},\zeta_a \mathcal{R})$$

$$-(R/\mathcal{R})^{\ell+u_b+1} A^{\ell+u_b}_{n_a+\Lambda-\ell}(\zeta_b R,\zeta_a R)]$$

$$+(-1)^{\ell_c+t_b} K_{\lambda+t_b}(\zeta_b \mathcal{R})$$

$$\times (R/\mathcal{R})^{\ell-n_a-\Lambda-1} \hat{A}^{\ell+u_b}_{n_a+\Lambda-\ell}(\zeta_b \mathcal{R}, \zeta_a \mathcal{R}) \Bigg) \Bigg\} \quad , \quad (R \geq \mathcal{R}) \quad . \qquad (20)$$

$$= \mathcal{S}^{\lambda\ell}_{b1}(\zeta_b \mathcal{R}) \, \mathcal{S}^{t\Lambda}_{c2}(\zeta_c R)$$

$$\times \Bigg\{ -(-1)^{u_b+u_c+\ell_b+\ell_c} I_{\lambda+t_b}(\zeta_b \mathcal{R}) I_{t+t_c}(\zeta_c R)$$

$$\times (R/\mathcal{R})^{\ell-n_a-u_c-1} \tilde{E}^{\ell+u_b,\Lambda+u_c}_{\Lambda+\ell+1-n_a}(\zeta_b \mathcal{R}, \zeta_c \mathcal{R}, \zeta_a \mathcal{R})$$

$$+(-1)^{t_b+\ell_c+u_c} K_{\lambda+t_b}(\zeta_b \mathcal{R}) I_{t+t_c}(\zeta_c R)$$

$$\times \Big[(R/\mathcal{R})^{\ell+u_b+1} \tilde{E}^{\Lambda+u_c,\ell+u_b}_{\Lambda+\ell+1-n_a}(\zeta_c R, \zeta_b R, \zeta_a R)$$

$$-(R/\mathcal{R})^{\ell-n_a-u_c-1} \tilde{E}^{\Lambda+u_c,\ell+u_b}_{\Lambda+\ell+1-n_a}(\zeta_c \mathcal{R}, \zeta_b \mathcal{R}, \zeta_a \mathcal{R}) \Big]$$

$$+(-1)^{t_b+t_c} K_{\lambda+t_b}(\zeta_b \mathcal{R}) K_{t+t_c}(\zeta_c R)$$

$$\times (R/\mathcal{R})^{\ell+u_b+1} \tilde{E}^{\ell+u_b,\Lambda+u_c}_{\Lambda+\ell+1-n_a}(\zeta_b R, \zeta_c R, \zeta_a R) \Bigg\}$$

$$+\mathcal{S}^{\lambda\ell}_{b1}(\zeta_b \mathcal{R}) \frac{(n_c+\ell_c+1)!}{(2\ell_c+1)!!} (\zeta_c R)^{-n_c-\ell_c-2}$$

$$\times \Bigg\{ \delta_{\Lambda,t+\ell_c} \frac{(2\Lambda-1)!!}{(2t+1)!!} \Big[(-1)^{\ell_b+u_b} I_{\lambda+t_b}(\zeta_b \mathcal{R})$$

$$x(R/\mathcal{R})^{\Lambda+\ell-n_a} {}^{\ell+u_b}_{a}E_{\Lambda+\ell+1-n_a}(\zeta_b\mathcal{R},\zeta_a\mathcal{R})$$

$$+(-1)^{\ell_b+t_b} K_{\lambda+t_b}(\zeta_b\mathcal{R}) \left((R/\mathcal{R})^{\Lambda+\ell-n_a} {}^{\ell+u_b}_{a}\tilde{E}_{\Lambda+\ell+1-n_a}(\zeta_b\mathcal{R},\zeta_a\mathcal{R}) \right.$$

$$\left. -(R/\mathcal{R})^{\ell+u_b+1} {}^{\ell+u_b}\tilde{E}_{\Lambda+\ell+1-n_a}(\zeta_b R,\zeta_a R) \right) \Big]$$

$$+\delta_{t,\Lambda+\ell_c} \frac{(2t-1)!!}{(2\Lambda+1)!!} (-1)^{\ell_c+t_b} K_{\lambda+t_b}(\zeta_b\mathcal{R})$$

$$x(R/\mathcal{R})^{\ell+u_b+1} {}^{\ell+u_b}\hat{A}_{n_a+\Lambda-\ell}(\zeta_b R,\zeta_a R) \Big\} , \quad (R\le\mathcal{R}) \quad . \tag{21}$$

with $\mathcal{S}^{\lambda\ell}_{b1}$ and $\mathcal{S}^{t\Lambda}_{c2}$ denoting the "operators",

$$\mathcal{S}^{\lambda\ell}_{b1}(\rho) = (-1)^{n_b-\ell_b} \sum_{u_b=0}^{n_b} \sum_{j_b=0}^{\min\{n_b-u_b,[(n_b-\ell_b)/2]\}} \sum_{s_b=0}^{n_b-u_b-j_b}$$
$$(t_b=n_b-j_b-s_b-u_b)$$

$$x \frac{(n_b-\ell_b)!}{(n_b-\ell_b-2j_b)!(2j_b)!!} \frac{(n_b-j_b)!}{s_b!t_b!u_b!} \frac{(\ell_b+\lambda+\ell+1)!!}{(\ell_b+\lambda+\ell+1-2s_b)!!} \rho^{1-j_b-s_b} , \tag{22}$$

$$\mathcal{S}^{t\Lambda}_{c2}(\rho) = (-1)^{n_c-\ell_c} \sum_{u_c=0}^{n_c} \sum_{j_c=0}^{\min\{n_c-u_c,[(n_c-\ell_c)/2]} \sum_{s_c=0}^{n_c-u_c-j_c}$$
$$(t_c=n_c-j_c-s_c-u_c)$$

$$x \frac{(n_c-\ell_c)!}{(n_c-\ell_c-2j_c)!(2j_c)!!} \frac{(n_c-j_c)!}{s_c!t_c!u_c!} \frac{(\ell_c+t+\Lambda-1)!!}{(\ell_c+t+\Lambda-1-2s_c)!!} \rho^{-1-j_c-s_c} \ . \tag{23}$$

IV. EVALUATION OF 2- AND 3-PARAMETER SPECIAL FUNCTIONS

All the special functions of Eqs. (15)-(19) can be expressed in closed form in terms of exponential-type integrals and inverse powers. The formulas follow along the lines of Eqs. (44)-(47) of Ref. 8 for the functions $A_n^{\ell}(x,y)$ and $\hat{A}_n^{\ell}(x,y)$. For instance, the finite sum for $E_n^{\ell_1\ell_2}(x,y,z)$ is

$$E_n^{\ell_1\ell_2}(x,y,z) = \sum_{\mu_1=0}^{\ell_1} \sum_{\mu_2=0}^{\ell_2} \frac{(\ell_1+\mu_1)!\ x^{-\mu_1-1}}{(\ell_1-\mu_1)!\ (2\mu_1)!!} \frac{(\ell_2+\mu_2)!\ y^{-\mu_2-1}}{(\ell_2-\mu_2)!(2\mu_2)!!}$$

$$\times\ E_{n-\ell_1+\mu_1-\ell_2+\mu_2}(x+y+z) \quad . \tag{24}$$

This equation is an excellent way to calculate $E_n^{\ell_1\ell_2}(x,y,z)$, because all terms have the same sign. In the corresponding equation for $\tilde{\mathbb{E}}_n^{\ell_1\ell_2}(x,y,z)$, there is a sign alternation, the consequence of which is numerical cancellation and unsuitability for numerical calculations. There is on the other hand an infinite expansion for $\tilde{\mathbb{E}}_n^{\ell_1\ell_2}(x,y,z)$ in which all the terms have the same sign and whose convergence is like that of the exponential function:

$$\tilde{\mathbb{E}}_n^{\ell_1\ell_2}(x,y,z) = \sum_{\nu_1=0}^{\infty} \sum_{\nu_2=0}^{\infty} \frac{x^{\ell_1+2\nu_1}\, y^{\ell_2+2\nu_2}\, \hat{\alpha}_{-n+2\ell_1+2\nu_1+2\ell_2+2\nu_2+2}(z)}{(2\ell_1+2\nu_1+1)!!(2\nu_1)!!(2\ell_2+2\nu_2+1)!!(2\nu_2)!!} . \tag{25}$$

Equation (25) is a numerically stable way to calculate $\tilde{\mathbb{E}}^{\ell_1\ell_2}(x,y,z)$. A similar equation with one finite and one infinite summation exists for $\tilde{E}_n^{\ell_1\ell_2}(x,y,z)$.

Each of the two- and three-parameter special functions has an integral representation [cf. Eqs. (42) and (43) of Ref. 8] from which recursion relations follow. For instance, the integral representation for $A_n^{\ell}(x,y)$ is

$$A_n^{\ell}(x,y) = \int_1^{\infty} dt\ t^{n+\ell+1} e^{-yt} K_{\ell}(xt) \quad , \tag{26}$$

and two recursion relations immediately follow from the recursion relations for spherical Bessel functions,

$$A_n^{\ell+1}(x,y) = A_{n+2}^{\ell-1}(x,y)+\frac{2\ell+1}{x} A_n^{\ell}(x,y) \quad , \tag{27}$$

$$(n+1)A_n^{\ell}(x,y) = yA_{n+1}^{\ell}(x,y)+xA_{n+2}^{\ell-1}(x,y)-e^{-y}K_{\ell-1}(x) \quad . \tag{28}$$

The first relation has no cancellations and is absolutely stable for positive values of x and y. Each two-parameter function has two linearly independent (ignoring relations changing only the inhomogeneous terms) recursion relations, while each three-parameter function has three. We shall discuss the recursion relations more completely in a future paper. In general, each special function has one recursion relation that is always stable, and others whose numerical stability depends on the values of the variables and parameters.

Armed with finite sums, infinite sums, and recursion relations for the special functions, it is then straightforward to write a computer program based on Eqs. (8), (20), and (21).

V. COMPUTER PROGRAMS

Two computer programs based on the above formulas have been written. The first evaluates each two- and three-parameter function using the appropriate (i.e., numerically stable) finite or infinite summation in terms of exponential-integral type functions. The ordinary special functions were calculated as described in Ref. 8. The n, ℓ, and m quantum numbers of the STO's may be chosen arbitrarily, provided that various dimensions are set sufficiently large. This first program was written in single precision.

The second computer program makes extensive use of stable recursion formulas to calculate the two- and three-parameter special functions. This second program is entirely in double precision. Nevertheless, because of its use of recursion relations, it executes considerably faster than the first program, and it is roughly comparable with the (mostly single precision) program POLYCAL kindly given to us by R. M. Stevens.

Restriction to absolutely stable recursion formulas, as opposed to the extensive use of conditionally stable ones, does result in some inefficiency. When calculating $A_n^{\ell}(x,y)$, for instance, the explicit formula for the cases $\ell=0$ and $\ell=1$ are very simple. The use of these values and the stable recursion formula (27) then can be used to generate all needed values, along with quite a few extra ones that are "intermediates." Figure 1 illustrates this "overhead," which is greater (but more difficult to illustrate in a figure) for the three-parameter functions. In a subsequent version of the second computer program,

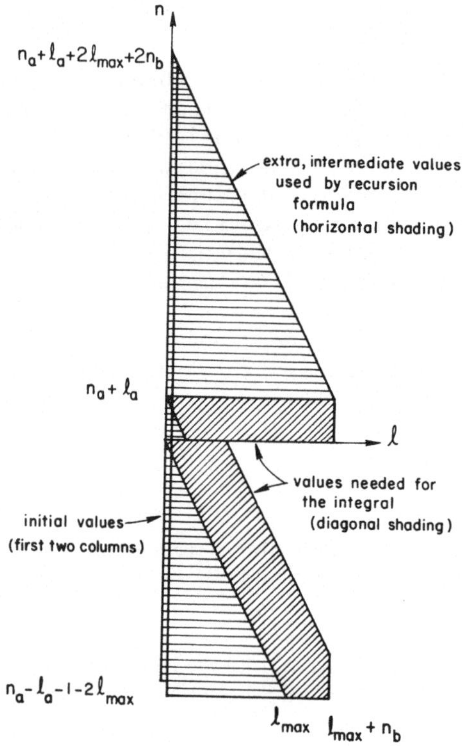

Figure 1. Diagram of parameter space for storage of $A_n^{\ell}(x,y)$ when calculated by the recursion formula $A_n^{\ell+1}=A_{n+2}^{\ell-1}+\frac{2\ell+1}{x}\,A_n^{\ell}$

conditionally stable recursion relations will be used to cut down on some of this overhead.

The current version of the second program does not take into operational account the ultimate accuracy to which the integral is desired. It calculates a large, fixed number of special functions, each to a prespecified accuracy, even if the final integral requires only a few, less accurate functions. A future version of the program will eliminate such "overkill."

Yet another consideration is asymptotics - both of the integral itself and of the individual terms $I^{t\ell\lambda\Lambda}_{c;ab}(R)$. When the internuclear distances are large, often there is a simple asymptotic formula that is sufficiently accurate and much faster to calculate. We touch on asymptotics in the next section and plan to implement such considerations in the final version of the computer program.

VI. ASYMPTOTICS AND CONVERGENCE

Equation (8) for the (1-2)-type three-center integral is an infinite expansion. The question naturally arises, how fast does it converge.[14,15] More precisely, what is the asymptotic form for the radial functions, as well as for all the special functions, as functions of all the relevant variables and parameters. The question is too big to answer definitively here: we shall be content with a few numerical examples and heuristic ideas.

Table 1 illustrates the convergence rate for several representative integrals, which are defined in Fig. 2. (The choice of s-orbitals is strictly to make the presentation of the examples manageable, since then $t=\ell=\lambda=\Lambda$, and the number of quantities to tabulate is a minimum.) Note that Integrals 1(a) and 1(b) are the same, except for how $\vec{R}$ and consequently $\vec{\mathcal{R}}$ are defined; simiarly for Integrals 2(a) and 2(b). It is apparent that convergence is slower when the ratio of R and $\mathcal{R}$ approaches unity.[14,15]

Also tabulated in Table 1 is a quantity labelled "asymptotic estimate" that we now explain. Suppose that the major contributions to the integral come from where the electrons are far apart. Then one should be able to "misuse" the Laplace expansion for $1/r_{12}$, taking always r_1 in the numerator and r_2 in the denominator and obtain a result that resembles the arbitrary irregular-solid-harmonic generalization of the one-electron three-center nuclear attraction integral,[9]

$$\int dV_1 \int dV_2 \; \psi_c^*(\vec{r}_1)\psi_a^*(\vec{r}_2-\vec{R})\psi_b(\vec{r}_2-\vec{R}-\vec{\mathcal{R}}) \quad r_{12}^{-1}$$

TABLE I. Convergence of (1-2)-type three-center integrals specified in Fig. 2.

ℓ	Contribution of Term ℓ	Asymptotic estimate	Contribution of Term ℓ	Asymptotic estimate
	Integral 1(a)		Integral 1(b)	
0	0.0032767377	0.0032842117	0.0021894669	0.0021894745
1	-14463639	-14650908	786689	786733
2	6366599	6591307	44705	44720
3	-2784432	-2986363	3489	3493
4	1208345	1361189	342	343
5	-520550	-623656	39	40
6	222933	287041	5	5
7	-95084	-132644	1	1
8	40462	61517		
9	-17205	-28622		
10	7320	13356		
11	-3119	-6249		
12	1332	2931		
13	-571	-1378		
14	245	649		
15	-106	-306		
16	46	145		
17	-20	-69		
18	9	33		
19	-4	-15		
Total	0.0022729938	0.0022730074	0.0022729939	0.0022730079
	Integral 2(a)		Integral 2(b)	
0	0.0007579428	0.0007583364	0.0006610472	0.0006610474
1	0	0	131548	131549
2	-991919	-999708	4091	4091
3	0	0	164	164
4	191895	198179	6	6
5	0	0		
6	-40166	-43766		
7	0	0		
8	8527	10174		
9	0	0		
10	-1795	-2438		
11	0	0		
12	372	596		
13	0	0		
14	-76	-147		
15	0	0		
16	15	37		
17	0	0		
18	-3	-9		
Total	0.0006746279	0.0006746281	0.0006746279	0.0006746283

TABLE I. Continued

ℓ	Contribution of Term ℓ	Asymptotic estimate	Contribution of Term ℓ	Asymptotic estimate
	Integral 3			
0	0.0001218275	0.0001223865		
1	58578	61405		
2	4375	5197		
3	432	648		
4	51	107		
5	7	22		
6	1	5		
7	0	1		
Total	0.0001281719	0.0001291251		

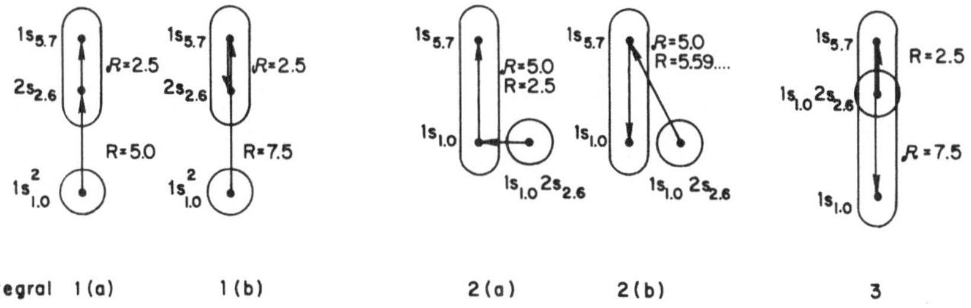

Figure 2. Geometry and orbital exponents for the five $(1s)^3 2s$ integrals of Table 1.

$$\sim (n_c+\ell_c+1)!\zeta_c^{-n_c-\ell_c-2} \int dV_2 r_2^{-\ell_c-1} Y_{\ell_c}^{m_c}(\theta_2,\phi_2)^* \psi_a^*(\vec{r}_2-\vec{R})\psi_b(\vec{r}_2-\vec{R}-\vec{R}). \quad (29)$$

Formulas for the three-center one-electron irregular-solid-harmonic integral[9] in general have two contributions: an "ordinary" part and a part that depends on how the conditionally convergent integral is evaluated near the origin when $\ell \geq 2$. Comparison of Eq. (64) of Ref. 9 with Eqs. (9) and (10) above reveal that the "ordinary" part of the one-electron integral is precisely the "Kronecker-delta" terms of Eqs. (9) and (10). In this sense, the Kronecker-delta terms, which are the ones explicitly involving the two-parameter special functions, hold as asymptotic behavior of the two-electron integral, and their contribution to the integral is listed separately in Table 1 in the "asymptotic estimate" columns. The accuracy of the estimate is remarkably good, even in some cases in which the two charge distributions are interpenetrating. Of course the time required to calculate the two-parameter-function contribution is much shorter than the time to calculate the complete integral, and considerable savings would result from not calculating the three-parameter functions when their contribution would be negligible.

VII. SUMMARY

The derivative operators that appear in Fourier-transform based formulas for the (1-2)-type three-center integral of $1/r_{12}$ with STO's have been removed from the formulas by a combination of explicit differentiation, folding into spherical Bessel functions, and definition of two- and three-parameter special functions. The evaluation of the new special functions by finite expansion, infinite expansion, and recursion has been sketched. Computer programs based on these formulas have been written, and data produced by the programs has been used to illustrate the convergence and asymptotic behavior of the integral. Next to be included in the program are conditionally stable recursion formulas, a consistent level of accuracy (vs. "overkill"), and asymptotics. The primary advantages of such a program are the generality with which the STO quantum numbers can be specified and the theoretical control on accuracy. It also appears that the final program will be acceptably fast.

REFERENCES

a) Present address: BDM Corporation, 7915 Jones Branch Drive, McLean, VA 22102.

1. Silverstone, H.J.: 1967, Proc. Natl. Acad. Sci. U.S. 58, p.34.

2. Silverstone, H.J.: 1968, J. Chem. Phys. 48, p.4098.

3. Silverstone, H.J.: 1968, J. Chem. Phys. 48, p.4106.

4. Silverstone, H.J. and Kay, K.G.: 1968, J. Chem. Phys. 48, p.4108.

5. Kay, K.G. and Silverstone, H.J.: 1969, J. Chem. Phys. 51, p.956.

6. Kay, K.G. and Silverstone, H.J.: 1969, J. Chem. Phys. 51, p.4287.

7. Kay. K.G. and Silverstone, H.J.: 1970, J. Chem. Phys. 53, p.4269.

8. Todd, H.D., Kay, K.G. and Silverstone, H.J.: 1970, J. Chem. Phys. 53, p.3951.

9. Silverstone, H.J. and Todd, H.D.: 1971, Int. J. Quantum Chem. 4, p. 371.

10. The two-center integral program is based on Ref. 8.

11. Asymptotic forumlas are used in the POLYCAL program of R.M. Stevens.

12. A current version of the (1-2)-type three-center integral program is described in the doctoral dissertation of Moats, R.K.: 1981, Johns Hopkins University, unpublished.

13. Abramowitz, M. and Stegun, I.A.: 1964, Handbook of Mathematical Functions, Appl. Math. Ser. No. 55, National Bureau of Standards, Wasnington, D.C..

14. Sahni, R.C. and LaBudde, C.D.: 1960, J. Chem. Phys. 33, p.1015.

15. LaBudde, C.D. and Sahni, R.C.: 1960, J. Chem. Phys. 33, p.1022.

TRANSLATION OF ANGULAR MOMENTUM EIGENFUNCTIONS USING NUMERICAL SPHERICAL BESSEL TRANSFORMS[1]

James D. Talman
Departments of Applied Mathematics and Physics, and Centre for Chemical Physics, University of Western Ontario, London, Ontario, Canada, N6A 5B9

The problem of carrying out the expansion in spherical harmonics of an angular momemtum eigenfunction is considered from a numerical viewpoint. The problem can be reduced to the calculation of spherical Bessel transforms, i.e., integral transforms with kernel $j_\ell(kr)$. A very effective method for carrying out such transforms numerically using logarithmic coordinates has recently been developed, and this method is applied to the translation problem. It is found that the results are not very satisfactory for $\ell=0$ wave functions because of the cusp at the origin. It is shown that the difficulty can be eliminated by removing the cusp using a counter term for which analytic results are known. A brief discussion of the possibility of applying the logarithmic coordinate spherical Bessel transform method to the multicenter integral problem is included.

INTRODUCTION

The problem of calculating multicenter integrals is related to the problem of expanding an angular momentum eigenfunction centered at one point in terms of angular momentum eigenfunctions centered at another point. That is, if

$$f(\underline{r}) = f_\ell(r)\, Y_{\ell m}(\hat{\underline{r}}), \tag{1}$$

can $f(\underline{r} - \underline{a})$ be expanded in terms of angular momentum eigenfunctions centered at the origin? Since $f(\underline{r} - \underline{a})$ depends on the directions of $\underline{r}$ and $\underline{a}$, it is possible to write

$$f(\underline{r} - \underline{a}) = \sum_{LML'M'} f_{LML'M'}(r)\, Y_{LM}(\hat{\underline{r}})\, Y_{L'M'}(\hat{\underline{a}}) \tag{2}$$

and the problem is to determine $f_{LML'M'}(r)$. (Through this article

C. A. Weatherford and H. W. Jones (eds.), ETO Multicenter Molecular Integrals, 91–101.

various functions derived from f_ℓ will also be denoted by f and distinguished by the nature of their indices.) If the solution is known, the multicenter integral problem is in principle, if not in practice, solved.

Much of the work on this problem has been directed toward finding analytic solutions in the case of Slater type orbitals: $f_\ell(r) = r^{\ell+s}\, e^{-\alpha r}$. In this note the problem will be considered as a purely numerical one in which the functions $f_\ell(r)$ and $f_{LML'M'}(r)$ are given or determined in tabular form. The method therefore applies to Slater type orbitals, but can be extended to numerical orbitals. It may be useful as well for problems in which the result could be in principle obtained analytically, but the algebra is intractible. For example, the translation properties of the product of two functions, $f_1(\underline{r} - \underline{a}_1)\; f_2(\underline{r} - \underline{a}_2)$ may be required.

A general expression for the function $f_{LML'M'}(r)$ in the case that $\underline{a}$ is in the direction of the z axis has been given by Sharma[2] and simplified by Rashid[3]. The result obtained is essentially for the function

$$f_{LM}(r) = \sum_{L'} f_{LML'0}(r) \tag{3}$$

The result involves polynominals in r and r^{-1} multiplying functions of the form

$$\int_{|r-a|}^{r+a} \rho^{2s+1}\, f_\ell(\rho)\, d\rho \tag{4}$$

and could be calculated either numerically or analytically. Some problem of cancellation errors, might however arise for large L and small r values, because the result behaves like r^L although the individual terms behave like r^{-L} in this limit.

FOURIER TRANSFORM METHOD

The method described in this work is based on the Fourier transform of $f(\underline{r})$:

$$f(\underline{r}) = (2\pi)^{-3} \int e^{-i\underline{k}\cdot\underline{r}}\, \tilde{f}\,(\underline{k})\, d\underline{k} \tag{5}$$

where

$$\begin{aligned} \tilde{f}(\underline{k}) &= \int e^{i\underline{k}\cdot\underline{r}} f(\underline{r})\; d\underline{r} \\ &= 4\pi\, i^{\ell}\; Y_{\ell m}\; (\hat{\underline{k}}) \int_0^\infty j_\ell(kr) f_\ell(r) r^2\; dr \end{aligned} \tag{6}$$

$$= 4\pi\, i^{\ell}\, \tilde{f}_{\ell}\,(k)\; Y_{\ell m}(\hat{\underline{k}}), \tag{7}$$

$$\tilde{f}_{\ell}\,(k) = \int_0^{\infty} j_{\ell}(kr)\; f_{\ell}(r)\; r^2\; dr. \tag{8}$$

$\tilde{f}_{\ell}\,(k)$ is called the ℓ spherical Bessel transform of f_{ℓ}.

We can then write

$$f(\underline{r} - \underline{a}) = (2\pi)^{-3} \int e^{i\underline{k}.\underline{a}}\; e^{-i\underline{k}.\underline{r}}\; \tilde{f}\;(\underline{k})\; d\underline{k}$$

$$= (2\pi)^{-3} \int \sum_{LML'M'} i^{L'-L}(2L+1)\;(2L'+1)$$

$$\times\; Y_{LM}(\hat{\underline{k}})^*\; Y_{LM}\;(\hat{\underline{r}})\; Y_{L'M'}(\hat{\underline{k}})^*\; Y_{L'M'}(\hat{\underline{a}})$$

$$\times\; 4\pi\; i^{\ell} Y_{\ell m}(\hat{\underline{k}})\; j_L(kr) j_{L'}(ka)\; \tilde{f}_{\ell}\;(k)\; d\underline{k} \tag{9}$$

(The spherical harmonics in (9) are $(4\pi/2\ell+1)^{1/2}$ times the usual spherical harmonics.) The integration over the direction of $\underline{k}$ in (9) gives

$$4\pi(-1)^m \begin{pmatrix} L & L' & \ell \\ 0 & 0 & 0 \end{pmatrix} \begin{pmatrix} L & L' & \ell \\ M & M' & -m \end{pmatrix}.$$

Comparison of (9) and (2) then yields

$$f_{LML'M'}(r) = (-1)^m \begin{pmatrix} L & L' & \ell \\ M & M' & -m \end{pmatrix} f_{LL'}(r) \tag{10}$$

where

$$f_{LL'}(r) = \frac{2}{\pi}\; i^{\ell+L'-L} (2L+1)\,(2L+1) \begin{pmatrix} L & L' & \ell \\ 0 & 0 & 0 \end{pmatrix}$$

$$\times \int_0^{\infty} j_L(kr)\; j_{L'}(ka)\; \tilde{f}_{\ell}(k)\; k^2\; dk \tag{11}$$

The problem of calculating $f_{LL'}(r)$ therefore reduces to calculating the ℓ spherical Bessel transform of f_{ℓ}, multiplying by $j_{L'}(ka)$ and calculating the L transform of the product.

LOGARITHMIC COORDINATE SPHERICAL BESSEL TRANSFORM METHOD

The calculation of integrals of the form of eq. (8) presents severe numerical difficulties in the case that k is large since the integral oscillates rapidly and a very fine mesh in the r variable is required. In a typical example, k_{max} and r_{max} might be 50 and 10, respectively. If $k_{max}\,\Delta r$ is of order 0.1, then $\Delta r < .02$ and therefore 5000 or more points in the r mesh are required. A similar number of points in the k mesh are required, so that the calculation is of order 5000^2 operations.

This problem can be avoided by changing variables in the transform to $\rho = \ell n\ r$ and $\kappa = \ell n\ k$[4,5]. It is convenient to define new functions $\hat{f}\ (\rho) = f_\ell(e^\rho)$ and $\hat{g}\ (\kappa) = \tilde{f}_\ell\ (e^\kappa)$.

Then

$$\hat{g}(\kappa) = \int_{-\infty}^{\infty} j_\ell(e^{\kappa+\rho})\ \hat{f}(\rho)\ e^{3\rho}\ d\rho. \tag{12}$$

This can be rewritten as

$$\hat{g}(\kappa) = e^{(m-3/2)\kappa} \int_{-\infty}^{\infty} e^{(3/2-m)(\kappa+\rho)} j_\ell(e^{\kappa+\rho}) \times e^{(3/2+m)\rho}\ \hat{f}(\rho)\ d\rho. \tag{13}$$

The integral in (13) can be recognized as a convolution; it is the integral on $(-\infty,\infty)$ of a function of $\kappa+\rho$ multiplying a function of ρ. The integral can therefore be expressed using Fourier transforms as

$$\hat{g}(\kappa) = (2\pi)^{-1}\ e^{(m-3/2)\kappa} \int_{-\infty}^{\infty} e^{i\kappa t}\ \phi(t)M(t)dt \tag{14}$$

where

$$\phi(t) = \int_{-\infty}^{\infty} e^{it\rho}\ e^{(m+3/2)\rho}\ f_\ell(e^\rho)d\rho. \tag{15}$$

$$M(t) = \int_{-\infty}^{\infty} e^{-it\rho}\ e^{(3/2-m)\rho}\ j_\ell(e^\rho)d\rho. \tag{16}$$

An analytic expression for $M(t)$ is given in Ref. 4.

The integer m in (13) - (15) can be chosen arbitrarily between 0 and ℓ; it represents the possibility of factoring various powers of r out of $j_\ell(kr)$ and including them in $f_\ell(r)$. It was shown in Ref. 3 however, that m=0 yields the best accuracy at large κ values, and $m=\ell$ yields the best accuracy at small κ ($\kappa \to -\infty$) values.

The method of calculation is to obtain the Fourier transforms of (14) and (15) numerically. The function M(t) is calculated analytically; this needs to be done only once. The numerical Fourier transforms can be performed using the fast Fourier transform algorithm for which the number of operations is proportional to $N\ \ell n\ N$, where N is the number of points in the ρ and κ meshes.

It was shown in some numerical examples in Ref. 4 that the method gives remarkable accuracy at large values of the transform variable when m in eqs. (13) - (15) is chosen to be 0. There is a loss of accuracy as $k \to 0$, $\kappa \to -\infty$, but this can be circumvented by straightforward means that are discussed below.

A more serious limitation, that has implications for the translation problem, was mentioned in Ref. 4. If the function $f_\ell(r)$, of eq. (8) has a discontinuity, or discontinuous derivative, at $r = R$, the transform function $\tilde{f}_\ell(k)$ has an oscillatory behavior, like $k^{-p}\, j_\ell(kR)$, for some p depending on the nature of the discontinuity. The mesh points in the k variable are spaced a distance $e^{\kappa}\Delta\kappa = k\Delta\kappa$ apart and for large k the oscillatory nature of the function is clearly not satisfactorily represented. Conversely, if the function to be transformed is oscillatory, difficulties will arise because of this inadequate description of the oscillatory behavior. This problem arises in the transform of eq. (11) because of the factor $j_L'(ka)$. This difficulty will be discussed further below.

COMMENTS ON THE NUMERICAL IMPLEMENTATION

The integrals in eqs. (14) and (15) are truncated to

$$\begin{aligned} I(\kappa) &= \int_{-\infty}^{\infty} e^{i\kappa t}\, \phi(t)M(t)dt \\ &\doteq \int_{-T}^{T} e^{i\kappa t}\, \phi(t)M(t)dt \\ &= 2\,\mathrm{Re} \int_{o}^{T} e^{i\kappa t}\, \phi(t)M(t)dt \end{aligned} \tag{17}$$

$$\begin{aligned} \phi(t) &= \int_{-\infty}^{\infty} e^{it\rho}\, h(\rho)d\rho \\ &\doteq \int_{\rho_{min}}^{\rho_{max}} e^{-it\rho}\, h(\rho)d\rho \end{aligned} \tag{18}$$

Mesh points $\kappa_{min} + m\Delta\kappa$, $n\Delta t$, $\rho_{min} + p\Delta\rho$ where, m, n, p = 0, 1, ..., N-1 are introduced. This leads to arrays I_m, ϕ_n, M_n, and h_p of the corresponding function values at the mesh points. The integrals are replaced by summations, so that

$$\begin{aligned} \phi_n &= \sum_{p=o}^{N-1} e^{in\Delta t(\rho_{min} + p\Delta\rho)} h_p\, \Delta\rho \\ &= e^{in\Delta t\rho_{min}} \sum_{p=o}^{N-1} e^{i\,\Delta\rho\,\Delta t\, np}\, h_p\, \Delta\rho \end{aligned} \tag{19}$$

$$I_m = 2\,\mathrm{Re} \sum_{n=o}^{N-1} e^{i\,\Delta\kappa\,\Delta t\, mn} e^{in\kappa_{min}\Delta t} \phi_n M_n \Delta t \tag{20}$$

The summations in (19) and (20) can be carried out using the fast Fourier transforms technique provided $\Delta\kappa$, Δt and $\Delta\rho$ are restricted by $\Delta\rho\Delta t = \Delta\kappa\Delta t = 2\pi/N$ and therefore necessarily, $\Delta\rho = \Delta\kappa$.

The quantities $e^{i(\kappa_{min} + \rho_{min})\Delta tn} M_n$ are calculated and stored the first time the program is called.

The value of $\phi(t)$ calculated numerically satisfy the identity $|\phi_{N-n}|=|\phi_n|$ and are therefore spurious for large t values. Therefore $\phi(t)$ is arbitrarily replaced by 0 for $t > T/2$, or $\phi_n = 0$, $n > N/2$.

The calculated values are inaccurate for k close to 0. This problem has been circumvented in the results presented as follows. For $\ell=0$ the analytic values of the function M(t) can be replaced by values obtained by calculating the integral in eq. (16) numerically. The result is then the same, apart from cancellation errors, as calculating the integral in eq. (8) directly numerically using the trapezoidal rule. This method is essentially that proposed by Siegman. Values of $\tilde{f}_\ell(k)$ were calculated by this method for k values such that $\tilde{f}_\ell(k)/\tilde{f}_\ell(0)$ 1/2.

For $\ell \neq 0$, $\tilde{f}_\ell(k)$ was extrapolated to k=0 using the k^ℓ dependence. The extrapolation is carried out from the point at which $\tilde{f}_\ell(k+\Delta k)/\tilde{f}_\ell(k)$ is closest to $((k+\Delta k)/k)^\ell = e^{\ell\Delta\kappa}$.

NUMERICAL RESULTS

In this section some numerical results from the application of the logarithmic coordinate method will be discussed. In Table I results for the $\ell=0$ transform of e^{-r} are given:

$$\int_0^\infty e^{-r} j_o(kr) r^2 dr = 2(1+k^2)^{-2} \tag{21}$$

Two sets of numerical results are given, calculated with N=64, $\Delta\rho=0.25$ and N=128, $\Delta\rho=0.125$. In each case $\rho_{min} = -12.0$. It would probably have been better to take a larger $\Delta\rho$ and a smaller ρ_{min} in the latter case. The accuracy for $\ell=0$ is strongly governed by ρ_{min} since the integrand in eq. (15) decreases like $e^{3\rho/2}$ for $\kappa \to -\infty$. For $\rho_{min} = -12$, this is about 10^{-8} and probably limits the accuracy obtained.

As another application of the method we show in Table II results for the convolution or overlap integral

$$C(R) = \int e^{-\alpha|\underline{r} - \underline{R}|} e^{-\beta r} d\underline{r} \tag{22}$$

calculated with $\alpha = 1.0$, $\beta = 0.5$. This integral is given by

$$C(R) = 8 \int_0^\infty j_o(kR)\ f(k)\ g(k)\ k^2 dk \tag{23}$$

where f(k) and g(k) are the $\ell=0$ transforms of $e^{-\alpha r}$ and $e^{-\beta r}$. The

Table I. Analytic and numerical values for the $\ell=0$ spherical Bessel transform of e^r, $2(1+k^2)^{-2}$, calculated with $\rho_{min} = -12$.

κ	k	analytic	N=64, $\Delta\rho=0.25$	N=128, $\Delta\rho=0.125$
-1.0	0.36788	1.5516070	1.5516070	1.5516070
0	1.0	0.5	0.49999996	0.50000000
1.0	2.71828	(-1).28418673	(-1).28418667	(-1).28418673
2.0	7.3891	(-3).64700750	(-3).64700677	(-3).64700757
3.0	20.0855	(-4).12227731	(-4).12227677	(-4).12227805
3.5	33.1154	(-5).16600286	(-5).16600512	(-5).16601037

Table II. Values of the convolution integral C(R) defined in eq. (22) calculated analytically and numerically.

R	analytic	N=64, $\Delta\rho=0.25$	N=128, $\Delta\rho=0.125$
0.36788	7.3641017	7.3641018	7.3641017
1.0	6.8837855	6.8837856	6.8837855
2.71828	4.5845143	4.5845177	4.5845143
7.3891	0.73363769	0.73363769	0.73363769
20.0855	(-2).16855775	(-2).16855905	(-2).16855776
33.1154	(-5).264683	(-5).264714	(-5).264698

spherical Bessel transform method is a remarkably simple, fast and accurate method to calculate such an integral.

Results for the function $f_{LL}(r)$ with $\ell=0$ defined by eq. (11) are given in Table III. In the results given $f_\ell(r) = e^{-r}$ and a=1. Results are given for L=0, 2 and 4.

It is seen that the results are quite unsatisfactory. The reason for this is that $\tilde{f}_\ell(k)$ falls off slowly, like k^{-4}, for increasing k and therefore large k values are important. However, the integrand contains the oscillatory factor $j_{L'}(ka)$ and the oscillations are not treated adequately in the logarithmic coordinates.

The k^{-4} behavior of $\tilde{f}(k)$ is caused by the cusp in f(r) at r=0. The problem could be reduced by subtracting from f(r) a function for which $f_{LL}(r)$ is known analytically so as to eliminate the cusp. The known analytic contribution can then be added back into the result. A Slater orbital is a convenient choice for the function to be subtracted out.

To illustrate that this is a feasible approach the functions $f_{LL}(r)$ have been calculated for

$$f(r) = e^{-r} - 2e^{-r/2}$$

for which $f'(0) = 0$. The results are given in Table IV and are seen to be greatly improved over the results for the function e^{-r}. This gives considerable support to the conjecture that the poor results for $f(r) = e^{-r}$ are caused by the slow rate of decrease of $\tilde{f}(k)$. The problem should be less severe at larger ℓ values, since $\tilde{f}_\ell(k)$ in general behaves like $k^{-\ell-4}$ at large k.

DISCUSSION

The logarithmic coordinate method of calculating spherical Bessel transforms may be useful for the multicentral integral problem in various ways. One application is to the calculation of $f_{LL'}(r)$ defined in eq. (11) for functions that are given purely numerically. It may be less useful in the case of Slater orbitals since the analytic result must be calculated in any event if the technique of removing the cusp analytically is used. It may be also useful, however, for linear combinations of Slater orbitals, such as occur in analytic atomic wave functions[6].

In calculating a multicenter integral an angular momentum expansion of a product

$$f(\underline{r} - \underline{R}_1)\, g(\underline{r} - \underline{R}_2)$$

about some intermediate center $\underline{R}_{12}$ may be generated, in the form

Table III. Values of the function $f_{LL}(r)$ defined in eq. (3) for $f(r) = e^{-r}$, L = 0, 2, 4. The factor $2/\pi$ is omitted.

r	analytic	N=64, $\Delta\rho$=0.25	N=128, $\Delta\rho$=0.125
		L = 0	
0.36788	0.5645634	0.564767	0.564573
1.0	0.46652191	0.466298	0.466533
2.71828	0.10778602	0.10777212	0.10778551
7.3891	(-2).109246	(-2).10910	(-2).109237
		L = 2	
0.36788	0.109764	0.11267	0.109829
1.0	0.469500	0.46758	0.46934
2.71828	0.1026734	0.10253	0.1026691
7.3891	(-3).86018	(-3).84301	(-3).85932
		L = 4	
0.36788	(-1).102629	(-1).09221	(-1).093470
1.0	0.22432	0.22157	0.22396
2.71828	(-1).166991	(-1).16481	(-1).166953
7.3891	(-4).60405	(-4).273	(-4).59128

Table IV. Values of the function $f_{LL}(r)$ defined in eq. (3) for $f(r) = e^{-r}-2e^{-r/2}$, for L = 0, 2, 4. The factor $2/\pi$ is omitted.

r	analytic	N=64, $\Delta\rho$=0.25	N=128, $\Delta\rho$=0.125
		L = 0	
.36788	-1.3085467	-1.3085476	-1.30855468
1.0	-1.1937540	-1.1937522	-1.1973541
2.71828	-0.68255020	-0.68256624	-.68255044
7.3891	(-1).78493954	(-1).78493955	(-1).78493954
20.0855	(-3).14125634	(-3).14125642	(-3).14125633
		L = 2	
.36788	-(-1).25899730	-(-1).25918669	-(-1).25899927
1.0	-0.12624149	-0.12623946	-0.12624153
2.71828	-0.14368360	-0.14368261	-0.14368359
7.3891	-(-1).17674061	-(-1).17673953	-(-1).17674060
20.0855	-(-4).28379103	-(-4).28368746	-(-4).28378791
		L = 4	
.36788	-(-3).73906172	-(-3).67256008	-(-3).73971157
1.0	-(-1).14972789	-(-1).14972129	-(-1).14972860
2.71828	-(-2).91742835	-(-2).91735419	-(-2).91742738
7.3891	-(-3).49832455	-(-3).49817956	-(-3).49832211
20.0855	-(-6).43309	-(-6).41145	-(-6).43260

$$\sum_{\lambda\mu} F_{\lambda\mu}(r)\, Y_{\lambda\mu}(\hat{\underline{r}})$$

where $\hat{\underline{r}}$ is the direction from $\underline{R}_{12}$ to $\underline{r}$. The functions $F_{\lambda\mu}(r)$ can be given analytically for Slater orbitals as infinite sums of products of the translated amplitudes for each of the factors. Further integrals on $\underline{r}$ may be done conveniently in momentum space, using $\tilde{F}_{\lambda\mu}(k)$ which can be readily obtained numerically.

In the same way a four center coulomb integral involving two products of the form above, leads to integrals of the form

$$\int_0^\infty j_L(kR)\, \tilde{F}_{\lambda\mu}(k)\, \tilde{G}_{\lambda'\mu'}(k)\,dk$$

where $R = |\underline{R}_{12} - \underline{R}_{34}|$. The logarithmic coordinate method may again be applicable to this integral.

REFERENCES

1. Supported by the National Science and Engineering Research Council of Canada.

2. Sharma, R.R.: 1976, Phys. Rev. A 13, p.517.

3. Rashid, M.A.: 1981, J. Math. Phys. 22, p.217.

4. Talman, J.D.: 1978, J. Comp. Phys. 29, p.35.

5. Siegman, A.E.: 1977, Optics Lett. 1, p.13.

6. Clementi, E. and Roetti: 1974, At. Data Nucl. Data Tables 14, p.177.

SPHERICAL-HARMONIC EXPANSION TECHNIQUES FOR MULTICENTER INTEGRALS OVER STO'S. A RE-EXAMINATION FOR VECTOR PROCESSING COMPUTERS.

H. H. Michels
United Technologies Research Center, East Hartford, Connecticut 06108

In multicenter integral calculations, it is useful to express an atomic orbital as a spherical-harmonic expansion about a point displaced from the orbital's center. Through judicious choice of these expansion points, it is possible to calculate the general electron-repulsion integral for orbitals displaced at most by distances of the order of the bond lengths.

We have re-examined the utility of these orbital expansions for STO's in light of the availability of new computers with vector processing hardware. In particular, we have carried out studies for STO's typical of atoms through Ne. Impressive times have been achieved by new programs implemented on the CRAY-1 computer where the products of orbital ℓ-expansions are essentially reduced to several multiplication cycle times. Convergence problems still remain for core-like STO's (large ζ's) which suggests that a different approach should be implemented to handle this class of integrals.

INTRODUCTION

One of the main difficulties in the application of any ab initio method to the study of the electronic structure of polyatomic systems is the necessity for rapid evaluation of the many electron-repulsion integrals that are required. Previous approaches to this problem have proceeded in two general directions.

The first is to employ Slater-type orbitals (STO's) as the primitive basis functions and to utilize various expansion methods for either an orbital or orbital product onto another center.[1-6] A somewhat different approach, based on a Gaussian integral transform, has been employed by Shavitt and Karplus.[7-9] An independent, but equivalent, integral transform approach based on convolution methods was first reported by Bonham, Peacher and Cox[10] and later refined by Monkhorst, et al.[11,12] Silverstone, et al.[13,14] discuss alternate transform approaches.

C. A. Weatherford and H. W. Jones (eds.), ETO Multicenter Molecular Integrals, 103–121.

The second general approach has been to directly employ Gaussian-type orbitals (GTO's) as the primitive basis functions.[15-19] The primary advantage of this approach is the great simplification in the mathematics since the product of two Gaussian orbitals located on separate centers can be written as a single Gaussian located at a point on the line between these centers. The primary disadvantage is the choice of basis functions with inherently improper long-range behavior, resulting in poorer wavefunction convergence. However, the severe computational problems associated with multicenter integrals over STO's has biased most ab initio molecular studies to the use of Gaussian orbitals as primitive basis functions. Their use is usually coupled with efficient schemes for contraction of a large number of GTO's to more representable forms for use in molecular calculations. Such contraction schemes include those proposed by Dunning and Hay[20] and by Pople and co-workers.[21]

Historically, the major difficulty with all direct methods for evaluating electron-repulsion integrals over STO's has been the excessive computational time per integral that is required and the observation that this computational time often scales like $\sim\ell^4$ for integrals over higher angular momentum orbitals. This scaling time, however, can often be made to be proportional to ℓ^2 if the computations are arranged in terms of the charge distributions resulting from STO orbital products.

In this paper we address the problem of lengthy computational time per time per integral by examining the benefits that may arise from the use of computers with vector processing hardware. Such computers include the CRAY-1, ILLIAC IV and smaller machines with array processor hardware. We shall examine the potential advantages of such computers in handling spherical-harmonic expansion methods for STO's, minimizing the amount of computer time by arranging the algebraic operations over the ℓ-expansions to conform to vectorizable forms. We first review our approach to the evaluation of the general multicenter electron-repulsion integral, using spherical-harmonic expansions of an STO onto a common coordinate system. We then analyze the amount of computer time that is required for the several operations involved in evaluating these integrals: orbital transformations, construction of charge distributions and final integral evaluation from pairs of charge distributions.

SPHERICAL-HARMONIC EXPANSIONS

Our general method for evaluating an electron-repulsion integral is to obtain the two charge distributions as spherical-harmonic expansions about a common center, not necessarily coincident with any orbital center, and to carry out the final integrations in this common axial system. The general electron repulsion integral can be written as

$$(ab|cd) = \int \Psi_a^*(1)\Psi_b(1) \frac{1}{r_{12}} \Psi_c^*(2)\Psi_d(2)d\tau_1 d\tau_2 \quad (1)$$

where Ψ_a designates an STO of the form

$$\Psi_a = r_a^{n_a-1} e^{-\delta_a r_a} P_{\ell_a}^{m_a}(\cos\theta_a)\cos(|m_a|\phi_a) \qquad m_a \geq 0$$

$$\Psi_a = r_a^{n_a-1} e^{-\delta_a r_a} P_{\ell_a}^{m_a}(\cos\theta_a)\sin(|m_a|\phi_a) \qquad m_a < 0 \quad (2)$$

centered at point A and where (r_a, θ_a, ϕ_a) are the spherical polar coordinates relative to A from a space-fixed axial frame. We let Ψ_b, Ψ_c, and Ψ_d be defined in an analogous manner. Since Ψ_a, Ψ_b, etc., are not necessarily centered on the common axial system chosen for the integrations, we must first consider expansions of these orbitals in spherical harmonics to a space-fixed (r, θ, ϕ) axial system. We expand Ψ_a and Ψ_b as

$$\Psi_a^* = \delta_a^{-n_a+1} \sum_{\sigma=-\ell_a}^{\ell_a} \sum_{j=|\sigma|}^{\infty} w_{j_a}^{-\sigma}(r)\, P_j^{\sigma}(\cos\bar{\theta}_a)\, e^{i\sigma\bar{\phi}_a}$$

$$\Psi_b = \delta_b^{-n_b+1} \sum_{\tau=-\ell_b}^{\ell_b} \sum_{k=|\tau|}^{\infty} w_{k_b}^{\tau}(r)\, P_k^{\tau}(\cos\bar{\theta}_b)\, e^{i\tau\bar{\phi}_b} \quad (3)$$

where $(r, \bar{\theta}_a, \bar{\phi}_a)$ and $(r, \bar{\theta}_b, \bar{\phi}_b)$ refer to the orientation of the axial frames for points A and B, respectively. Methods for evaluating the expansion coefficients, $W_j^{\sigma}(r)$, using recursion formulae, have been previously reported.[4,5]

These expansions can be viewed as resulting from a series of well-defined coordinate transformations. In Fig. 1, we indicate an initial coordinate system for ψ_a and ψ_b, centered at A and B, respectively, with arbitrary axial orientations. Figure 2 illustrates the coordinate system after orbital rotations which point the polar axis toward the common expansion point, P. This transformation is accomplished using the basic relationship

$$P_\ell^m(\cos\theta)\, e^{im\phi} = \sum_{\sigma=-\ell}^{\ell} D_\ell^{m\sigma}(\alpha, \beta, \gamma)\, P_\ell^{\sigma}(\cos\bar{\theta})\, e^{i\sigma\bar{\phi}} \quad (4)$$

where $D_\ell^{m\sigma}(\alpha, \beta, \gamma)$ are the group rotation coefficients corresponding to a transformation described by Eulerian angles (α, β, γ). A method for the rapid evaluation of these coefficients has been reported by Harris.[22]

The spherical harmonic expansion of an STO onto a point P can be written as

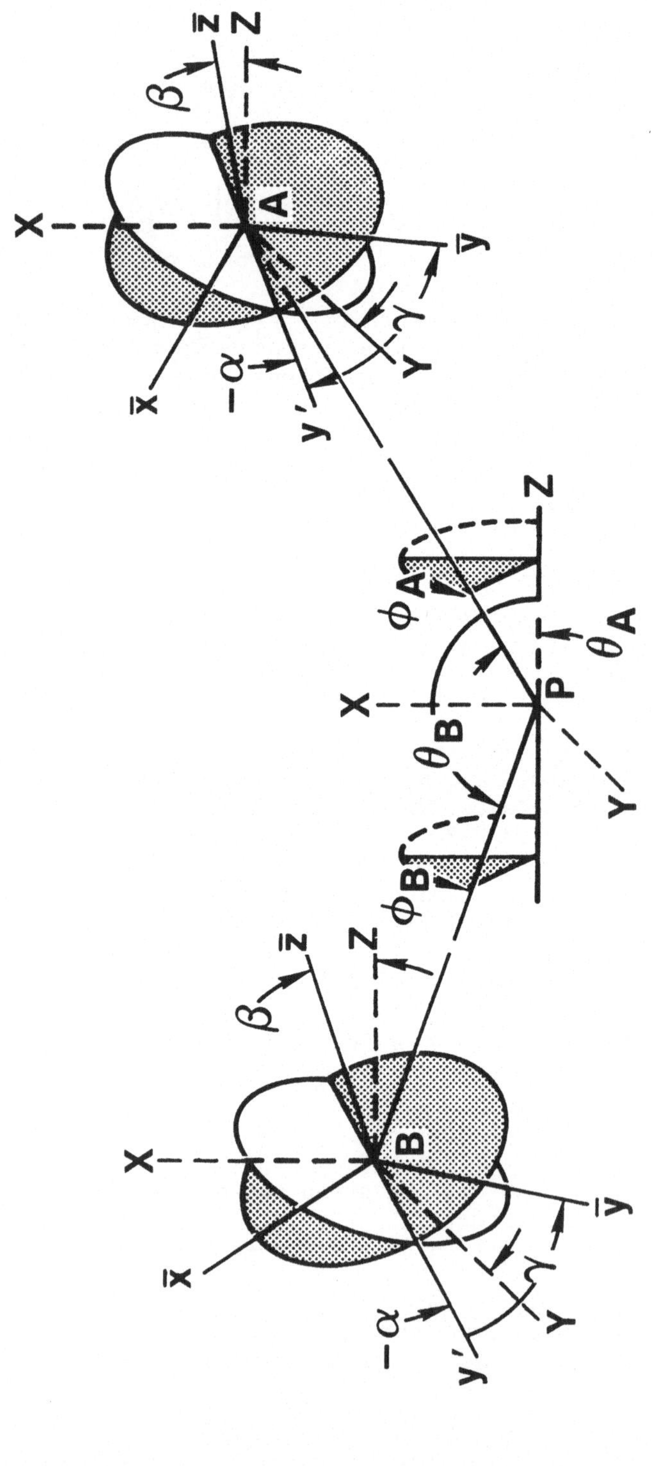

FIG. 1. INITIAL COORDINATE SYSTEM

FIG. 2. COORDINATE SYSTEM AFTER ROTATIONAL TRANSFORMATION

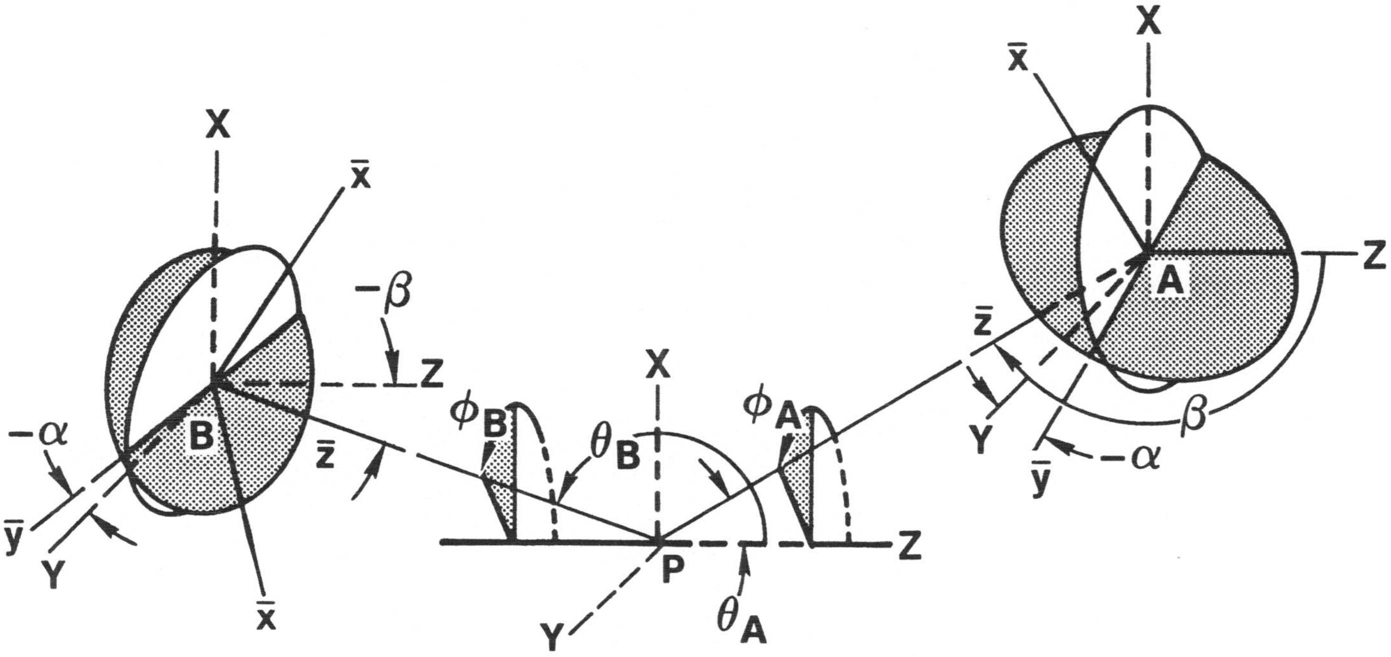

Rotations (α, β, γ) carry ψ_a to an axial frame where the new z-axis points toward the expansion point P. A similar rotation is performed on ψ_b.

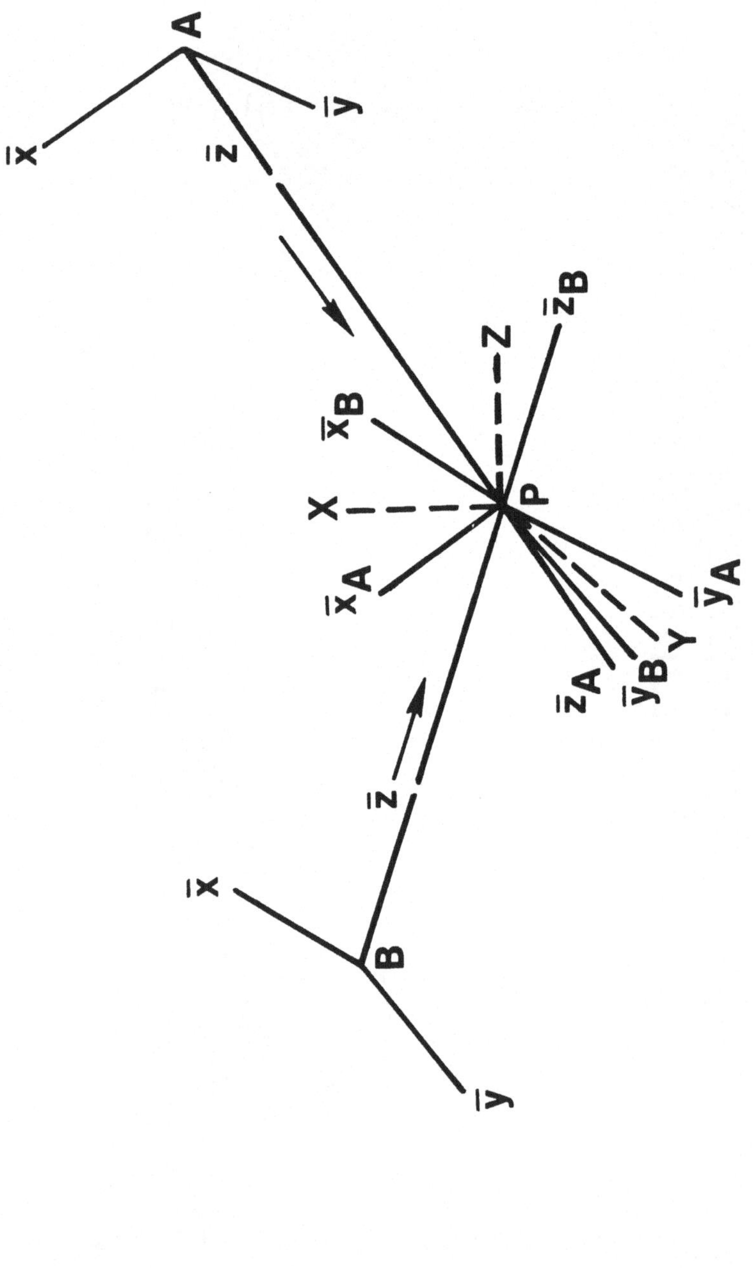

FIG. 3. TRANSLATIONAL TRANSFORMATION USING SPHERICAL HARMONIC EXPANSION

FIG. 4. ROTATIONAL TRANSFORMATIONS

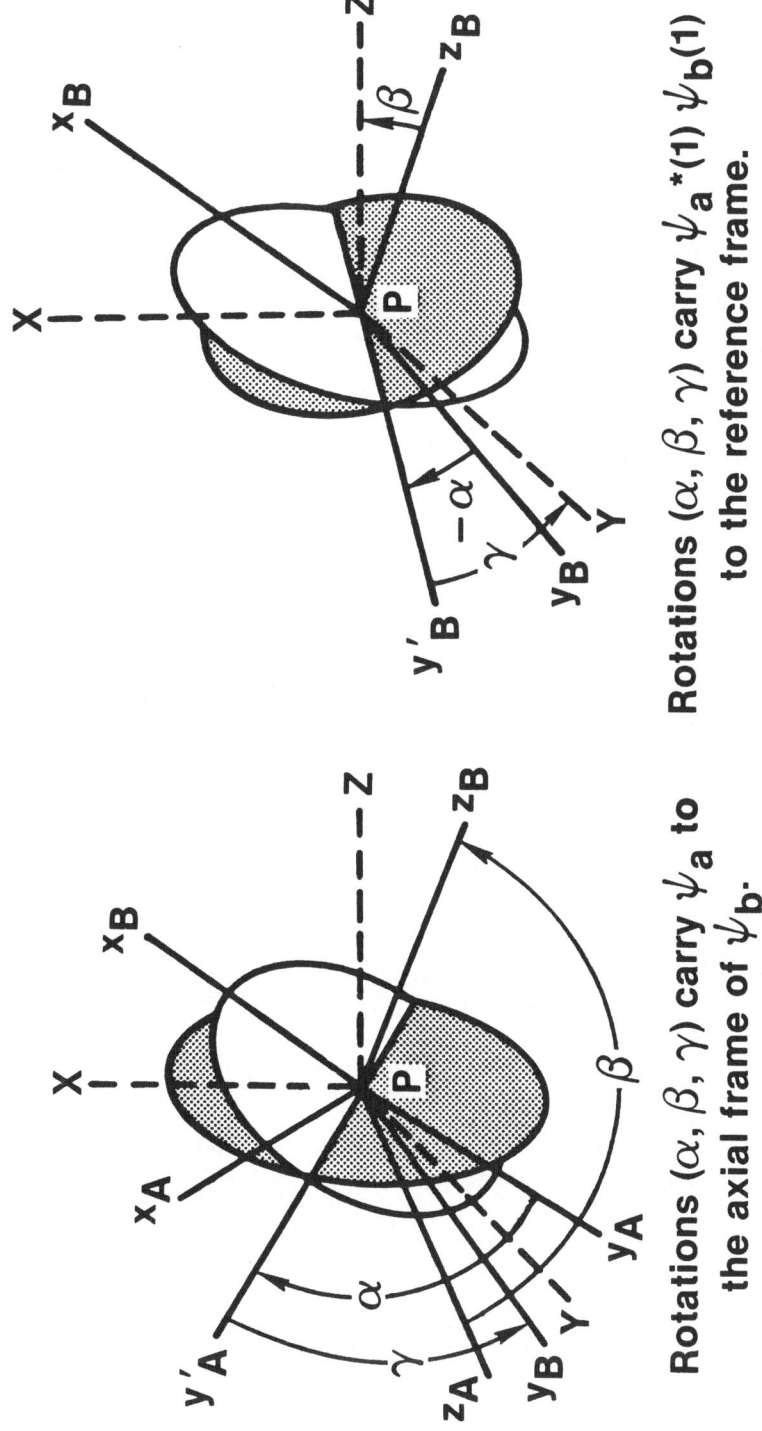

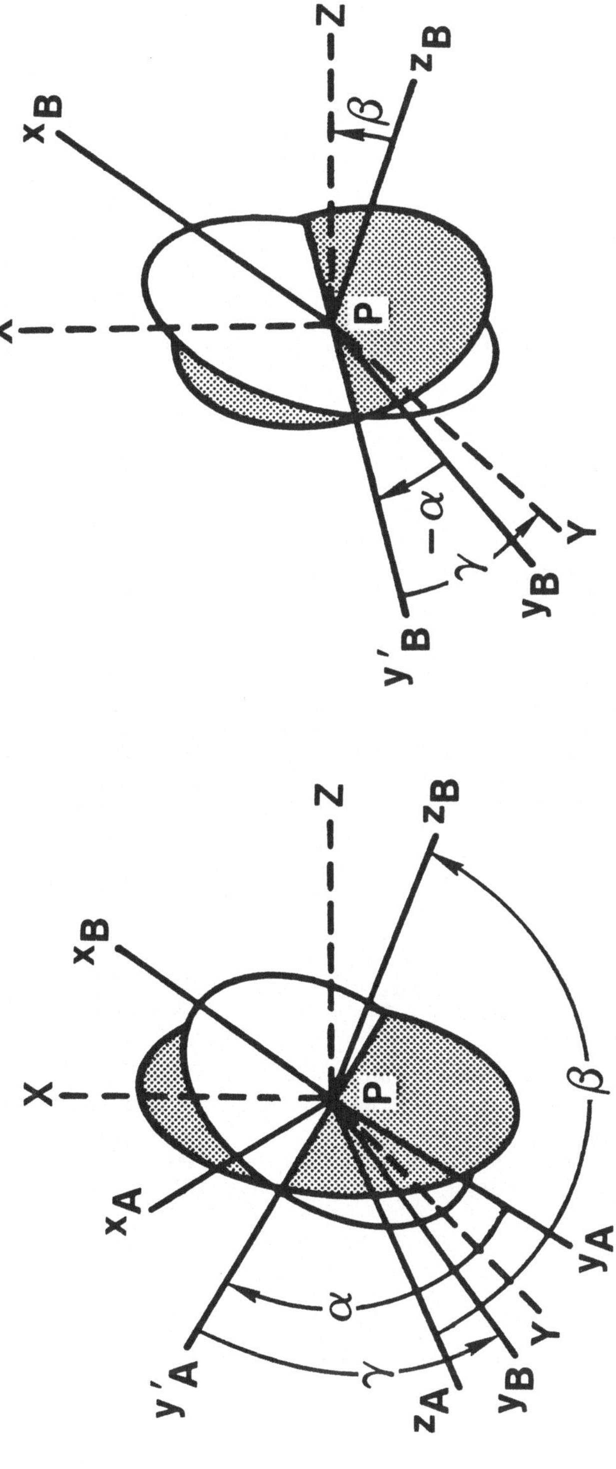

Rotations (α, β, γ) carry ψ_a to the axial frame of ψ_b.

Rotations (α, β, γ) carry $\psi_a{}^*(1)\,\psi_b(1)$ to the reference frame.

$$\Psi_a = r_a^{n_a-1} e^{-\delta_a r_a} P_{\ell_a}^{\sigma}(\cos\theta_a) e^{i\sigma\phi_a}$$

$$= \delta_a^{-n_a+1} \sum_{j=|\sigma|}^{\infty} V_{n_a, \ell_a, j}^{\sigma}(\delta_a r, \delta_a R) P_j^{\sigma}(\cos\bar{\theta}_a) e^{-i\sigma\phi_a} \quad (5)$$

where $V_{n_a,\ell_a,j}^{\sigma}$ are expansion coefficients that can be obtained either recursively or by direct methods.[4-6] These coefficients are similar to the ζ-function expansion coefficients described by Barnett and Coulson[3] and to the α-function discussed by Löwdin,[23] for which Sharma[24] has given closed formula expressions. In Fig. 3, a spherical harmonic expansion of Ψ_a and Ψ_b onto point P is illustrated. The representation of orbitals Ψ_a and Ψ_b is now that given by Eq. (3).

Following a rotation at point P which brings the coordinate system for Ψ_a coincident with that for Ψ_b or vice-versa, we can express the charge distribution corresponding to the product of two STO's as

$$\Psi_a^*(1)\, \Psi_b(1) = \sum_{\ell=0}^{\infty} \sum_{m=-\ell}^{\ell} f_{\ell}^{m}(r_1) P_{\ell}^{m}(\cos\theta_1) e^{im\phi_1} \quad (6)$$

where

$$f_{\ell}^{m}(r_1) = \delta_a^{-n_a+1}\, \delta_b^{-n_b+1} \quad (7)$$

$$* \sum_{j=0}^{\infty} \sum_{k=0}^{\infty} \sum_{\sigma=-<\ell_a,\ell_b}^{>\ell_a,\ell_b} C_{\ell,i,j,k}^{\sigma,m+\sigma}\, W_j^{-\sigma}(r_1)\, W_k^{m+\sigma}(r_2)$$

and $C_{\ell,j,k}^{\sigma,m+\sigma}$ are Clebsch-Gordan coefficients that can be generated most conveniently by recursion formulae. A similar analysis for the second charge distribution yields

$$\Psi_c^*(2)\Psi_d(2) = \sum_{\ell=0}^{\infty} \sum_{m=-\ell}^{\ell} g_{\ell}^{m}(r_2)\, P_{\ell}^{m}(\cos\theta_2)\, e^{im\phi_2} \quad (8)$$

where $g_{\ell}^{m}(r_2)$ is defined in a manner analogous to Eq. (7). The remaining step required for evaluation of the general multicenter integral given by Eq. (1) is a rotation of the charge distributions given by Eqs. (6) and (8) to a standard reference orientation. These rotational transformations are illustrated in Fig. 4.

INTEGRAL REDUCTION TO ONE-CENTER FORM

Our final reduction of Eq. (1) is to invoke use of the Green's function expansion for r_{12}^{-1} in spherical coordinates:

$$\frac{1}{r_{12}} = \sum_{\ell=0}^{\infty} \sum_{m=-\ell}^{\ell} \frac{(\ell-|m|)!}{(\ell+|m|)!} \frac{r_<^{\ell}}{r_>^{\ell+1}} P_\ell^m(\cos\theta_1) P_\ell^m(\cos\theta_2) e^{im(\phi_2-\phi_1)} \quad (9)$$

where $r_<$ and $r_>$ stand for the smaller and greater of r_1 and r_2, respectively. Substituting Eq. (9) into Eq. (1) and carrying out the indicated angular integrations yields:

$$(ab|cd) = \sum_{\ell=0}^{\infty} \sum_{m=-\ell}^{\ell} (2\ell+1)\frac{(\ell-|m|)!}{(\ell+|m|)!} \int_0^\infty dz z^{-2\ell-2} F_\ell^{-m}(z) G_\ell^m(z) \quad (10)$$

where

$$F_\ell^m(z) = \left(\frac{4\pi}{2\ell+1}\right)\frac{(\ell+|m|)!}{(\ell-|m|)!} \int_0^z dr r^{\ell+2} f_\ell^m(r_1) \quad (11)$$

and $f_\ell^m(r_1)$ is defined by Eq. (7). In a similar fashion, $G_\ell^m(z)$ is defined using $g_\ell^m(r_2)$ for the second charge distribution. In Eq. (10), use has been made of the symmetric integral transformation[5]

$$\int_0^\infty dr_1 \int_0^\infty dr_2 \frac{r_<^\ell}{r_>^{\ell+1}} f(r_1) g(r_2)$$

$$= (2\ell+1)\int_0^\infty dz z^{-2\ell-2} \int_0^z dr_1 r_1^\ell f(r_1) \int_0^z dr_2 r_2^\ell g(r_2) \quad (12)$$

The final integral given by Eq. (10) is to be evaluated numerically choosing points z_k and corresponding weights w_k. The inner integrations given by Eq. (11) and a similar form for $G_\ell^m(z)$ are also carried out numerically using a point-to-point integration formula in the mesh corresponding to the points, z_k. All other steps in the integral evaluation are carried out analytically.

As has been previously pointed out[6], expansions similar to Eq. (3) need be carried out only once for each orbital, expansions for charge distributions as given in Eq. (6) need be done N(N+1)/2 times for an N-orbital problem and only the final integration shown in Eq. (10) needs to be done for each different integral. Since we shall see that this last step can be performed very efficiently on vector processing machines, our average integral evaluation time may be significantly reduced by this symmetric reduction of the electron-repulsion integral to a charge distribution oriented form.

NUMERICAL METHODS

The computation form of Eq. (10) can be written as

$$(ab|cd) = \sum_{k=1}^{K} \sum_{\ell=0}^{\infty} \sum_{m=-\ell}^{\ell} W_k \left[(2\ell+1) \frac{(\ell-|m|)!}{(\ell+|m|)!} z_k^{-2\ell-2} \right]$$

$$\times F_\ell^{-m}(z_k)\, G_\ell^{m}(z_k) \tag{13}$$

This can be cast in the symmetric form

$$(ab|cd) = \sum_{k=1}^{K} \sum_{\ell=0}^{\infty} \sum_{m=-\ell}^{\ell} Q_{\ell,k}^{-m}(ab)\, Q_{\ell,k}^{m}(cd) \tag{14}$$

where,

$$Q_{\ell,k}^{m}(ab) = \frac{4\pi}{z_k^{\ell+1}} \left[\frac{W_k(\ell+|m|)!}{(2\ell+1)(\ell-|m|)!} \right]^{1/2} \int_0^{z_k} dr\, r^{k+2} f_\ell^{m}(r) \tag{15}$$

is a point-to-point integral for charge distribution (ab) and $Q_{\ell,k}^{m}(cd)$ is similarly defined. Several numerical problems arise in the evaluation of Eq. (15) owing to the complicated behavior of the integrand. Although the functions $f_\ell^{m}(r)$ are continuous, there are discontinuities in these functions at points corresponding to the location of the individual orbitals relative to our common expansion coordinate framework. A more serious problem is that the functions change sign several times in the r-range that must be integrated. This dictates that a fine mesh quadratic or cubic point-to-point integration be used in order to accurately follow these functions through sign changes. Higher order polynomials, used in an average representation sense, cannot be expected to accurately represent the nodal character of the functions, $f_\ell^{m}(r)$.

In our original study of this problem,[5,6] this point-to-point integration was carried out using cubic spline interpolation of the integrand. The end points were chosen to correspond to a 34-point Lobatto quadrature formula that was used in evaluating the final two-electron integral. A re-examination of this integration scheme indicated that this method was not entirely satisfactory for charge distributions arising from highly noded orbitals. A more accurate (∿6 significant figures) evaluation of Eq. (15) can be achieved by using a quadratic interpolating point-to-point integration between the outer quadrature points, z_k. This requires that the function, $f_\ell^{m}(r)$, be evaluated at 65 separate r-values and for the range of ℓ-values chosen for the spherical harmonic expansions. By proper storage of the functions, $f_\ell^{m}(r)$, Eq. (15) can be conveniently cast into a vectorizable form for the indicated integration. The outer integration, indicated in computational form by Eq. (14), can be cast as a dot product for efficient vector processing.

A careful analysis of all storage requirements is crucial for optimum vector processing of the various operations leading to the final integral evaluation. A first concern is the practical extent of the spherical harmonic expansion given by Eq. (5) for an orbital translation. Although numerical tests can be made to determine the number of terms required in Eq. (5) for a presdribed accuracy of the translated orbital, some maximum limit must be established for this summation for optimum vector processing and storage. After numerous numerical tests, an upper limit of 24 terms was chosen which ensures 5-figure accuracy for $\delta R \lesssim 8$. The various operations dependent upon this expansion limit, which have substantial storage requirements, are outlined below.

Orbital Rotation Coefficients

The transformation for axial orbital rotation is indicated in Eq. (4) where $D_{\ell}^{m\sigma}(\alpha,\beta,\gamma)$ are group rotation coefficients that must be evaluated for each orbital or charge distribution rotation. We have set the limits for our STO's at $n \leq 7$, $\ell \leq 3$, $|m_{\ell}| \leq \ell$, with the result that the maximum D-coefficient array has dimensions

$$D_{\ell}^{m\sigma} = \mathrm{DROT}(\ell,m,\sigma) = \mathrm{DROT}(7,7,24) \tag{16}$$

This results in a maximum of 1176 coefficients which can easily be held in central memory and efficiently generated as needed for each orbital or charge distribution.

Translation Coefficients

The translational expansion indicated in Eq. (5) is sufficiently complicated that some plan to store the coefficients, $V_{n,\ell,j}^{\sigma}(\delta r,\delta R)$, is indicated. We choose to evaluate these coefficients recursively[4], but alternate approaches in which these coefficients are evaluated directly[24,25] would not change the indicated storage requirements. For a maximum expansion length of 24 terms we have for a given STO with prescribed quantum numbers, (n,ℓ,σ) defined at 65 spatial points:

$$V_{n,\ell,j}^{\sigma} = \mathrm{VGEN}(r_i,j) = \mathrm{VGEN}(65,24) \tag{17}$$

Since we will need to rotate this orbital to the various axial frameworks of other orbitals in order to form change distributions, we will actually need to evaluate these coefficients for all $|\sigma| \leq j$ rather than for the original STO magnetic quantum number, σ. Since we now need $(2j+1)$ coefficients for each j, this extends the storage requirements of the V-coefficients to

$$\mathrm{VGEN}(r_i,k) = \mathrm{VGEN}(65,24^2) \tag{18}$$

with the second subscript indexed in the order $(\sigma = -j, -j+1, \ldots. +j-1, +j)$ for each successive j. This arrangement allows for more efficient vector processing of these coefficient arrays.

Each orbital thus requires a maximum of 37440 storage locations. For a very large core machine, such as a CRAY-2, it is practical to keep up to $\sim$40 orbitals in central memory with the result that very efficient mergings of these arrays to form charge distributions can be accomplished. For smaller core machines, these V-coefficients can be written in serial fashion onto a disc or other high capacity external storage device and read back into central memory as needed. Since the I-O time for reading these arrays is usually much greater than that required for merging these arrays to form charge distributions, a significant penalty is paid for their external storage.

Clebsch-Gordan Coefficients

The merging of orbitals to form tables of charge distributions is indicated in Eq. (7). Some strategies must obviously be formulated for handling the large number of Clebsch-Gordan coefficients that are required. These coefficients can be generated rapidly and conveniently using recursion formulae[6] but the indicated vector operations in Eq. (7) dictate their most convenient storage as a multi-dimensional array. Our choice is

$$C^{\sigma,m+\sigma}_{\ell,j,k} = C(j,k,\ell,\sigma) = C(24,24,24,7) \qquad (19)$$

for a maximum ℓ-expansion of 24 terms. This results in a central memory requirement of 96768 storage locations which is easily accommodated by a CRAY-size machine but may pose a problem for smaller core machines. In such cases, following the strategy for vectorizing the indicated matrix multiplications, we could write out the C-coefficients in separate blocks of 13825 elements for each σ value. We note that many of these coefficients are zero by symmetry but that the recursion relationships used for their generation dictate that the entire C-array be constructed. Further study of the storage of these coefficients and their use in Eq. (7) is clearly indicated for core-limiting machines where they cannot be kept in central memory.

Charge Distributions

As pointed out above, the indicated matrix multiplications in Eq. (7) are easily carried out in vectorized form provided the C-coefficients are stored as multi-dimensioned arrays. These indicated operations yield tables, $\delta_\ell^m(r)$, describing the various charge distributions formed from orbital products. The dimension of these tables is clearly

$$\delta^m_\ell(r) = \mathrm{TAB}(r_i,k) = \mathrm{TAB}(65,24^2) \qquad (20)$$

where the (ℓ,m) expansion is stored in the same vector fashion as for orbitals. It can be seen that at least three arrays (two for orbitals and one for the charge distribution), must be kept in central memory simultaneously. Since a pairwise merging of all orbitals is required, an optimum strategy is to keep as many orbitals as possible in central memory (for example, LMAX) and to merge all combinations to form charge distributions. This can then be followed by a serial merge of all orbitals in central memory with each remaining orbital that must be read in from an external storage device. This logic results in complete use of the first LMAX orbitals which can then be replaced in central memory from the remainder that are stored externally. Since the number of charge distributions that are formed is typically much larger than can be conveniently stored in the central memory of any computer, including a CRAY-2, the best strategy is to integrate each distribution according to Eq. (11) and write out the array to an external device. Since our indicated mesh for the integrated distribution is 32 points in z, we have

$$F_{\ell}^{m}(z_i) = \text{TABLE}(z_i,k) = \text{TABLE}(32,24^2) \tag{21}$$

where again the (ℓ,m) expansion is stored as a single array for simplicity in vector multiplication. For computational convenience in calculating two-electron integrals, the integrated charge distributions are stored as the symmetric form given by Eq. (15).

Two-Electron Integrals

The final numerical steps involve the pairwise merging of all charge distributions to produce two-electron integrals according to Eq. (14). Since the tabular data representing a charge distribution can be stored as a single vector if cast in the symmetric form shown in Eq. (15), the final calculation of an integral can be performed from a simple dot product. This can be accomplished using an intrinsic vector function on the CRAY or an equivalent vectorizable loop for other computers with array processor hardware. The large time penalty paid for I-O dictates that as many charge distributions as possible should be kept in central memory. This logic must be compromised with the fact that, for a problem involving a large number of STO's, the number of final two-electron integrals becomes large and it may be most convenient to keep these final integrals in central memory. On very large core machines such as the CRAY-1, it is still only possible to store $\sim$100 charge distributions in central memory and the final integral evaluationtime becomes I-O dependent for basis sizes greater than about 12 orbitals.

COMPUTATIONAL DETAILS

The focus of this work has been to reexamine spherical-harmonic expansion techniques for evaluating multicenter integrals over STO's in light of the availability of computers with vector processing hardware. This study involved both an analysis of algorithms and computer program structure, as discussed above, and timing runs for quantitative analysis of the cost benefits that can result from this new computer architecture. The timings of the several steps involved in the general spherical-harmonic expansion technique described above has been carried out for a CRAY-1 computer system.

The architecture of the CRAY-1 computer has been described in detail in the literature.[26] Basically, the system is a pipeline processor with separate pipes for memory addressing, multiplication, addition, etc. (This design contrasts with that of ILLIAC IV which was a true parallel processor machine but with very limited storage per processor.) There are a maximum of 64 elements in any register or operation pipe and therefore, vector operations of length 64 or less, after some start-up overhead, can be carried out in a single step. Longer vectors must be segmented into blocks of 64 or less, with some loss of efficiency. The CRAY-1 can execute 160 million floating point operations per second (160 MFLOPS) simultaneously on all vector registers and processors, but more typical speeds are about 80 MFLOPS owing to start-up overhead and data storage delays.

The program timing was initially carried out in an extensive fashion since an early goal was to discover rate determining steps, analyze the algorithm that was employed, and to then decide if a significant program coding modification was warranted. The CRAY FORTRAN compiler typically only attempts to vectorize the innermost FORTRAN DO-loop, and then only if certain restrictive conditions are met. Many of these restrictions can be overcome by a simple restructuring of the code. The final timings were carried out for the following separate operations: orbital transformations, construction of charge distributions and integral calculation, assuming the availability of charge distributions stored in central memory.

In Table I, we give typical computation times that have been achieved on a CRAY-1 system. A first observation is that the vectorizing FORTRAN compiler, as implemented on the CRAY, yields a substantial reduction in the required computer time for all operations. Since the theoretical maximum improvement using vector operations on the CRAY-1 is 64, we see that both the orbital transformation and final integral evaluation times are dramatically reduced using vector operations.

A second observation from Table I is that the time required for calculating a single charge distribution defines the rate determining step in our analysis. This result is similar to that found in our previous studies.[5,6] A significant further reduction in the time required for computing charge distributions can be achieved by optimization of the several algorithms leading to the result expressed by Eq. (15). The most noteworthy change here is to keep all of the Clebsch-Gordan coefficients in central memory that are required for forming charge distributions from orbital products. A factor of two or so improvement in the final integral time can be achieved through use of efficient dot product subroutines.

Finally, we note that, for higher angular quantum number STO's, the orbital transformation and charge distribution times scale by $\sim t_o*(\ell+1)$, where t_o is the time required for operations involving s-type orbitals. This result can be understood from Eq. (7) which

TABLE I

COMPUTATION TIMES FOR MULTICENTER INTEGRALS OVER STO'S

	TIME (SEC) CRAY-1		
4-CENTER INTEGRAL	ORBITAL TRANSFORMATION	CHARGE DISTRIBUTION	INTEGRAL TIME
(ss/ss) Vectors off	.172	.381	.0486
(ss/ss) Vectors on	.015	.067	.0024
(ss/ss) Vectors on-optimized code	.015	.025	.0010
(dd/dd) Vectors on	.042	.234	.0024
(dd/dd) Vectors on-optimized code	.042	.088	.0010

TABLE II

COMPUTATION TIMES AS A FUNCTION OF NUMBER OF STO'S

			4-CENTER (ss/ss) INTEGRALS			
				TIME (SEC) CRAY-1		
NORB	NCD	NTWOEL	ORBITAL TRANSFORM	CHARGE DIST.	TWOEL INTEGRALS	TIME/INTEGRAL
10	55	1540	.15*	36.90	3.63	.0264
			.15**	13.84	1.58	.0101
20	210	22155	.30	140.91	52.29	.00873
			.30	52.84	22.82	.00343
30	465	108345	.45	312.01	255.69	.00524
			.45	117.00	111.59	.00211
40	820	336610	.60	550.22	794.40	.00400
			.60	206.33	346.71	.00164
50	1275	813450	.75	855.52	1919.74	.00371
			.75	320.82	837.85	.00143

* Vectors on
**Vectors on + program optimization

indicates a linear dependence on the total summation lengths with quantum number, σ.

In Table II, we illustrate the time advantage to be gained from a charge distribution oriented approach to the calculation of a large number of multicenter electron repulsion integrals. For these timing runs, it was not possible to keep all of the charge distributions in central memory at the same time and the overhead of the I-O required for reading from external disc storage is reflected in the data.

We note that the orbital transformation time is always insignificant when a group of orbitals is considered. This results from the observation that this operation scales like N for an N-orbital problem and that the unit time/orbital is small ($\sim$15 msec). For many problems, we see that the rate determining step is the construction of charge distributions, an operation that scales like $N(N+1)/2$ for an N-orbital problem. Only for problems involving 30 orbitals or more does the computation time for final integral evaluation become rate controlling. Thus, even though the unit time for merging pairs of charge distributions is small, this operation must be performed $N(N+1)(N^2+N+2)/8$ times in an N-orbital problem and eventually becomes rate controlling in the total time for all integral operations.

A final observation from Table II is that, even for a 50 orbital problem, the I-O overhead in handling charge distributions destroys much of the benefit that is gained from vectorization. For this case, we realize average computation times per charge distribution of 250 msec, whereas the data in Table I indicate an in-core average time of 25 msec. This I-O overhead is not as severe for the process of final integral evaluation. Here we achieve computation times close to our computer limit of 1 msec per integral.

CONCLUSION

It is clear from the above discussion that spherical-harmonic expansion techniques can be implemented with high computational efficiency on computers with vector processing hardware. In particular, factors of 10-20 improvement in computation time are easily achievable on a system such as the CRAY-1. These factors are larger than the nominal improvement to be gained by program vectorization owing to the natural way that the spherical-harmonic ℓ-expansion can be treated on parallel processing or pipeline computer systems. There are, however, certain significant deficiencies of spherical-harmonic expansion techniques that limit the general applicability of the method. A summary of these advantages and disadvantages is given below.

Advantages

1. The applicability to STO's with high principle quantum numbers, such as $4f_{+3}$, is accomplished with relative ease. Orbital transform and charge distribution calculation times scale by $t^*(\ell+1)$ and $tt^*(\ell+1)$, respectively, where t and tt represent the unit times for these operations using s-type STO's. This is to be contrasted with the typical Gaussian orbital analysis which scales at best by $(\ell+1)^2$.

2. For the calculation of a large number of multicenter integrals, such as would be required for a typical molecular problem, the final integral computation time is mainly dependent on the time requirement of merging charge distributions and is independent of orbital ℓ-value.

3. The average time per integral decreases rapidly with the number of integrals calculated. The best time with current programs on the CRAY-1 system is $\sim$1 msec/integral.

4. The general uniformity of approach is logistically desirable since a proliferation of integral nomenclature and specialized evaluation procedures is avoided. As a correlary, the amount of computer programming required is minimized and the chances for program error are reduced.

Disadvantages

1. Despite the natural applicability of vector processing computers to the ℓ-expansion arrays, the method suffers from the relatively slow convergence of the ℓ-expansion for large values of the product of the STO screening constant times the displaced distance, δR. An ℓ-expansion of 24 terms is required for five figure accuracy for $\delta R \gtrsim 8$. This problem cannot easily be circumvented by arbitrarily increasing the maximum number of terms in the ℓ-expansion owing to the storage requirement problem to be discussed below.

2. The storage requirement per orbital scales like the number of integration points times the ℓ-expansion length, (NPT*LMAX). However, the storage requirement per charge distribution scales like NPT*LMAX^2, and since a minimum of three such distributions are required simultaneously in central memory, a practical limit for the ℓ-expansion is $\sim$50 terms. For an expansion of this extent, the individual charge distribution evaluation time increases to 1-2 seconds on a CRAY-1
system and the total integral computation times begins to approach 1
system and the total integral computation times begins to approach 1 hour for a molecular problem with $\sim$40 STO's.

3. A third disadvantage is that vector processing computer equipment seems to be almost a requirement for the method to show any competitive edge over a Gaussian orbital analysis, considering the very high speeds by which multicenter integrals can be calculated with such basis functions (1000/sec on the CYBER 76 system).

A final comment is that a reformulation of the expansions required for the general multicenter integral might prove to be fruitful. A general approach might be to expand the Fourier transform of a two-center charge distribution about some point chosen for optimum expansion convergence properties.[12] The general bipolar expansion[27] could then be employed to decouple the integrations into charge distribution quantities. Such an arrangement keeps the computational rate-controlling step at the charge distribution level, thereby gaining the advantage of scaling by $\sim N^2$ instead of by $\sim N^4$ for an N-orbital molecular problem. More detailed analysis is clearly needed in this area.

ACKNOWLEDGMENTS

The author wishes to acknowledge valuable discussions with Frank E. Harris who was instrumental in the initial development of the computational methods described here. The assistance of R. H. Hobbs and J. B. Addison with the development of the CRAY programs is acknowledged with pleasure. Use of the computing facilities at the Kirtland Air Force Base in Albuquerque is also acknowledged. This research was supported in part by AFOSR under Contrast F49620-81-C-0097.

REFERENCES

1. Coolidge, A.S.: 1932, Phys. Rev. 42, p.189.

2. Landshoff, R.: 1936, Z. Physik 102, p.201.

3. Barnett, M.P. and Coulson, C.A.: 1951, Phil. Tran. Soc. (London) 243, p.221.

4. Harris, F.E. and Michels, H.H.: 1965, J. Chem. Phys. 43, p.165.

5. Harris, F.E. and Michels, H.H.: 1966, J. Chem. Phys. 45, p.116.

6. Harris, F.E. and Michels, H.H.: 1967, Adv. Chem. Phys. 13, Prigogine, I. (Ed.), Interscience.

7. Shavitt, I. and Karplus, M.: 1962, J. Chem. Phys. 36, p.550.

8. Shavitt, I.: 1963, Meth. in Compt. Physics 2, B. Adler, et. al. (Eds.), Academic Press.

9. Shavitt, I. and Karplus, M.: 1965, J. Chem. Phys. 43, p.398.

10. Bonham, R.A., Peacher, J.L. and Cox, H.L.: 1964, J. Chem. Phys. 40, p.3083.

11. Monkhorst, H.J. and Harris, F.E.: 1972, Int. J. Quantum Chem. 6, p.601.

12. Graovac, A., Monkhorst, H.J. and Zivkovic, T.: 1973, Int. J. Quantum Chem. 7, p.233.

13. Silverstone, H.J.: 1968, J. Chem. Phys. 48, p.4098,4106,4108; 1969, J. Chem. Phys. 51, p.956,4287; 1970, J. Chem. Phys. 53, p.4169.

14. Silverstone, H.J. and Kay, K.G.: 1969, J. Chem. Phys. 50, p.5045.

15. Boys, S.F.: 1950, Proc. Roy. Soc. A (London) 200, p.542.

16. Boys, S.F.: 1960, Proc. Roy. Soc. A (London) 258, p.402.

17. Browne, J.C. and Poshusta, R.D.: 1962, J. Chem. Phys. 36, p.1933.

18. Harris, F.E.: 1963, Rev. Mod. Phys. 35, p.558.

19. Krauss, M.: 1963, J. Chem. Phys. 38, p.564.

20. Dunning, Jr., T.H. and Hay, P.J.: 1977, Methods of Electronic Structure Theory III, Schaefer, H.F. (Ed.), Plenum Press.

21. Krishnan, R., Binkley, J.S. and Pople, J.A.: 1980, J. Chem. Phys. 72, p.650. (Also see previous papers in this series.)

22. Harris, F.E.: 1963, Rev. Mod. Phys. 35, p.558.

23. Löwdin, P.O.: 1956, Adv. Phys. 5, p.1.

24. Sharma, R.R.: 1976, Phys. Rev. A 13, p.517.

25. Jones, H.W. and Weatherford, C.A.: 1978, Int. J. Quantum Chem. Symp. 12, p.483.

26. Johnson, P.: Feb., 1978, Comp. Design, 89.

27. Ruedenberg, K.: 1967, Theor. Chim. Acta 7, p.359.

ONE-CENTER ELECTRON REPULSION INTEGRALS FOR SLATER AND GAUSSIAN ORBITALS

Russell M. Pitzer
Department of Chemistry, Ohio State University, Columbus, OH 43210

One-center electron repulsion integrals can always be evaluated analytically,[1,2] if desired, whether Slater or Gaussian orbitals are used, but the formulas can be put in a number of algebraic forms. While making some modifications in an atomic self-consistent-field program[3], I derived integral formulas in an improved form which I had not seen elsewhere.

The integrals, once the angular integrals have been performed, can be written for Slater orbital radial functions of the form $r^{n-1}e^{-zr}$ as

$$R_\ell = \int_0^\infty\!\!\int r_1^{n_{pq}} e^{-z_{pq}r_1} \frac{r_<^\ell}{r_>^{\ell+1}} r_2^{n_{rs}} e^{-z_{rs}r_2} dr_1 dr_2$$

where

$$n_{pq} = n_p + n_q \qquad z_{pq} = z_p + z_q$$

$$r_< = \text{minimum of } r_1, r_2 \qquad r_> = \text{maximum of } r_1, r_2$$

and p,q,r,s index the principal quantum numbers and orbital exponents of the four radial functions.

The desirable characteristics of the final formula are (1) ease of computation and programming, (2) freedom from differencing effects, most simply accomplished by having all terms of the same sign, and (3) ability to save intermediate quantities to simplify the calculation of integrals with the same radial functions but other ℓ values

The integral can be rewritten as

C. A. Weatherford and H. W. Jones (eds.), ETO Multicenter Molecular Integrals, 123–127.

$$R_\ell = \int_0^\infty dr_1 r_1^{n_{pq}+\ell} e^{-z_{pq} r_1} \int_{r_1}^\infty dr_2 r_2^{n_{rs}-\ell-1} e^{-z_{rs} r_2} +$$

$$\times \int_0^\infty dr_2 r_2^{n_{rs}+\ell} e^{-z_{rs} r_2} \int_{r_2}^\infty dr_1 r_1^{n_{pq}-\ell-1} e^{-z_{pq} r_1}$$

Carrying out the integrations as indicated yields

$$R_\ell = \frac{(n_{pq}+\ell)!\,(n_{rs}-\ell-1)!}{z_{pqrs}^{\,n_{pqrs}-1}} \sum_{i=o}^{n_{rs}-\ell-1} \binom{n_{pq}+\ell+i}{i} \left(z_{pq}+z_{rs}\right)^{n_{rs}-\ell-i-2} z_{rs}^{\,i-n_{pq}+\ell}$$

$$+ \frac{(n_{rs}+\ell)!\,(n_{pq}-\ell-1)!}{z_{pqrs}^{\,n_{pqrs}-1}} \sum_{i=o}^{n_{pq}-\ell-1} \binom{n_{rs}+\ell+i}{i} \left(z_{pq}+z_{rs}\right)^{n_{pq}-\ell-i-2} z_{pq}^{\,i-n_{pq}+\ell}$$

where

$$z_{pqrs} = z_{pq} + z_{rs} \qquad\qquad n_{pqrs} = n_{pq} + n_{rs}$$

Combining the last terms of each sum, and expanding the powers of $(z_{pq} + z_{rs})$ gives

$$R_\ell = \frac{(n_{pqrs}-1)!}{z_{pqrs}^{\,n_{pqrs}-1}} \left\{ \frac{1}{z_{pq} z_{rs}} \right.$$

$$+ \binom{n_{pqrs}-1}{n_{rs}-\ell-1}^{-1} \sum_{i=o}^{n_{rs}-\ell-2} \binom{n_{rs}+\ell+i}{i} \sum_{j=o}^{n_{rs}-\ell-i-2} \binom{n_{rs}-\ell-i-2}{j} \frac{z_{pq}^{j}}{z_{rs}^{j+2}}$$

$$\left. + \binom{n_{pqrs}-1}{n_{pq}-\ -1}^{-1} \sum_{i=o}^{n_{pq}-\ell-2} \binom{n_{pq}+\ell+i}{i} \sum_{j=o}^{n_{pq}-\ell-i-2} \binom{n_{pq}-\ell-i-2}{j} \frac{z_{rs}^{j}}{z_{pq}^{j+2}} \right\}$$

Next, the order of the i and j sums is reversed, and standard properties of binomial coefficients[4] are used to evaluate and i sums, giving

$$R_\ell = \frac{(n_{pqrs}-1)!}{z_{pq} z_{rs} z_{pqrs}^{\,n_{pqrs}-1}} \left\{ 1 + \binom{n_{pqrs}-1}{n_{rs}-\ell-1}^{-1} \sum_{j=o}^{n_{rs}-\ell-2} \binom{n_{pqrs}-1}{n_{rs}-\ell-2-j} \left[\frac{z_{pq}}{z_{rs}}\right]^{j+1} + \binom{n_{pqrs}-1}{n_{pq}-\ell-1}^{-1} \sum_{j=o}^{n_{pq}-\ell-2} \binom{n_{pqrs}-1}{n_{pq}-\ell-2-j} \left[\frac{z_{rs}}{z_{pq}}\right]^{j+i} \right\}$$

Performing the sums in the opposite order and other rearrangements yields

$$R_\ell = \frac{(n_{pqrs}-1)!}{z_{pq} z_{rs} z_{pqrs}^{\,n_{pqrs}-1}} \left\{ 1 + \left[\sum_{i=o}^{n_{rs}-\ell-2} \binom{n_{pqrs}-1}{i} \left[\frac{z_{rs}}{z_{pq}}\right]^{i} \right] \Bigg/ \left[\binom{n_{pqrs}-1}{n_{rs}-\ell-1} \left[\frac{z_{rs}}{z_{pq}}\right]^{n_{rs}-\ell-1} \right] + \left[\sum_{i=o}^{n_{pq}-\ell-2} \binom{n_{pqrs}-1}{i} \left[\frac{z_{pq}}{z_{rs}}\right]^{i} \right] \Bigg/ \left[\binom{n_{pqrs}-1}{n_{pq}-\ell-1} \left[\frac{z_{pq}}{z_{rs}}\right]^{n_{pq}-\ell-1} \right] \right\}$$

The formula is now in the form of partial binomial expansion sums divided by the first terms omitted from the sums. Note that increasing the ℓ value gives an expression which contains a smaller number of the same quantities already computed. For normalized orbitals, a factor of the form $\sqrt{(2z)^{2n+1}/(2n)!}$ must be included for each of the four radial functions.

This expression of R_ℓ is only slightly simpler than the one given by Roothaan and Bagus[1] in that the order of the binomial coefficients is one smaller and there is one less term to be added up. Both expressions possess all the advantages listed earlier.

I found that the use of recursion relations was the most convenient way to evaluate these integral formulas with the defintion

$$E_k^n(x) = \left[\sum_{j=o}^{k-1} \binom{n}{j} x^j\right] \Big/ \left[\binom{n}{k} x^k\right]$$

The desired relation is

$$E_k^n(n) = k \left[1 + E_{k-1}^n (x)\right] \Big/ \left[(n - k + 1)x\right]$$

which is used with a starting value of $E_o^n(x) = 0$

The integral expression is now

$$R_\ell = \frac{(n_{pqrs}-1)!}{z_{pq} z_{rs} z_{pqrs}^{\;n_{pqrs}-1}} \left\{1 + E_{n_{rs}-\ell-1}^{n_{pqrs}-1} \left(\frac{z_{rs}}{z_{pq}}\right) + E_{n_{pq}-\ell-1}^{n_{pqrs}-1} \left(\frac{z_{pq}}{z_{rs}}\right)\right\}$$

Two sets of E's are computed and stored, first with $x = z_{rs}/z_{pq}$ and $\ell = n_{rs}-1$ to ℓ_{min}, then with $x = z_{pq}/z_{rs}$ and $\ell = n_{pq}-1$ to ℓ_{min} where ℓ_{min} is the lowest ℓ value needed. The R_ℓ can then easily be assembled for all the desired ℓ values.

The integral formulas for Gaussian orbitals, which have radial functions $r^{n-1}e^{-zr^2}$, can be derived by exactly the same approach. Only the Gaussian orbitals useful for molecular computations[5] were considered, so there is a restriction on the n, ℓ quantum number that n-ℓ be a positive integer.

The integer expression is

$$R_\ell = \frac{2(n_{pqrs}-3)!!}{\pi^{\frac{1}{2}} z_{pq} z_{rs} z_{pq}^{\;\frac{1}{2}(n_{pqrs}-3)}} \left\{1 + E_{\frac{1}{2}(n_{rs}-\ell-2)}^{\frac{1}{2}(n_{pqrs}-3)} \left(\frac{z_{rs}}{z_{pq}}\right) + E_{\frac{1}{2}(n_{pq}-\ell-2)}^{\frac{1}{2}(n_{pqrs}-3)} \left(\frac{z_{pq}}{z_{rs}}\right)\right\}$$

where $(2n-1)!!$ denotes the odd number factorial $1\cdot3\cdots(2n-1)$. The E functions are computed first with $x = z_{rs}/z_{pq}$ and $\ell = n_{rs}-2$ to ℓ_{min} in steps of 2, and then with $x = z_{pq}/z_{rs}$ and $\ell = n_{pq}-2$ to ℓ_{min} in steps of 2. The R_ℓ are then assembled for the desired ℓ values. A factor of $\sqrt{(2z)^{n+\frac{1}{2}}/(2n-1)!!}$ must be included for each of the four radial functions in order to obtain integrals for normalized orbitals.

This formula for Gaussian orbitals affords some computational simplification over that given by Huzinaga.[2]

REFERENCES

1. Roothaan, C.C.J. and Bagus, P.S.: 1963, Math. Comp. Phys. 2, p.47.

2. Huzinaga, S.: 1965, J. Chem. Phys. 42, p.1293.

3. Roos, B., Salez, C., Veillard, A. and Clementi, E.: 1968, IBM Research RJ518.

4. Korn, G.A. and Korn, T.M.: 1968, Mathematical Handbook for Scientists and Engineers, 2nd ed., McGraw-Hill, New York, Table 21.5-1(a).

5. Dunning, T.H. and Hay, P.J.: 1977, Methods of Electronic Structure Theory III, Schaefer, H.F. (Ed.), Plenum, New York, p.l.

A NUMERICAL STUDY ON THE CHOICE OF BASIS SETS USED FOR TRANSLATING ETOs IN MULTI-CENTER LCAO CALCULATIONS

B.L. Jain
Physics Department, Florida A&M University, Tallahassee, Florida 32307

1. INTRODUCTION

It is well recognized that exponential-type orbitals (ETOs) defined in Eq.(1) are the natural one-center spherical functions to represent molecules in bound-state as well as in scattering calculation because they exhibit the proper long and short range behavior to satisfy boundary conditions. However, calculations based on ETOs have proven to be rather difficult and time comsuming in comparison to other methods such as the one involving Gaussian-type functions (GTFs). An ETO has the form

$$\eta_{n\ell m}(\alpha,\vec{r}) = N_n f_{n\ell}(r)\, e^{-\alpha r}\, Y_\ell^m(\theta,\phi) \tag{1}$$

In order to make ETO based molecular calculations somewhat more time-efficient and tractible there has been a renewed interest in obtaining analytic series expansions of multi-center molecular integrals over ETOs[1-8]. In this regard, ETOs that exhibit radial as well as angular orthonormality have been exploited. This was done first by, Guseinov[5], and then by Weatherford and Jain[6], and by Filter and Steinborn[8]. All three of these works have used the same basis sets - namely the orthonormal functions first discussed by Slater [9]. The ETOs used by Guseinov are given in Eq.(2)

$$G_{n\ell m}(\alpha,\vec{r}) = G_{n\ell}(\alpha,r)\, Y_\ell^m(\theta,\phi)$$

$$G_{n\ell}(\alpha,r) = \left[\frac{(2\alpha)^3(n-\ell-1)!}{(n+\ell+1)!}\right]^{\frac{1}{2}} \sum_{\sigma=o}^{n-\ell-1} \frac{(-1)^\sigma\,(n+\ +1)!^{\frac{1}{2}}\,(2\alpha)^{\ell+\sigma}\, r^{\ell+\sigma}\, e^{-\alpha r}}{(n-\ell-1)!\,(2\ell+2+\sigma)!\,\sigma\,!} \tag{2}$$

An important characteristic of the Guseinov functions, called Orthonormal Slaters (ONS) from here on, is that the screening constant is arbitrary. As a result of this and the orthonormality, all ETO

C. A. Weatherford and H. W. Jones (eds.), ETO Multicenter Molecular Integrals, 129–133.

multi-center integrals can be reduced to sums over equal screening constant, two-center overlaps. Guseinov does his overlaps in elliptical co-ordinates. Filter and Steinborn[2], however have given a spherical coordinate, computable form for these overlaps. Filter and Steinborn's method was recently used by Weatherford and Jain[6] to evaluate multi-center, electron-molecule interaction integrals.

In addition to the three ETOs mentioned above, Jain and Weatherford[7] have added a fourth ETO - namely Hydrogenic Sturmians given by Eq.(3). Sturmians were first introduced to atomic and molecular physics by Rotenberg[10]. However, they seem to have escaped serious exploitation by atomic physicists until very recently. Sturmians are solutions of the hydrogen-atom Schrödinger equation for fixed energy, the lowest Sturmian being the exact 1s hydrogen wavefunction. One attractive feature of the Sturmians revelant to the molecule problem is that they also form a complete, orthonormal set (without adding continuum separately). The Hydrogenic Sturmians are given by

$$S_{n\ell m}(\gamma,\vec{r}) = S_{n\ell}(\gamma,r)\, Y_{\ell}^{m}(\theta,\phi)$$

$$S_{n\ell}(\gamma,r) = \left[\frac{4\gamma^2(n-\ell+1)!}{n(\ell+1)^3(n+\ell)!}\right]^{\frac{1}{2}} \sum_{\sigma=o}^{n-\ell-1} \frac{(-1)^{\sigma}(n+\ell)!\left(\frac{2\gamma}{\ell+1}\right)^{\sigma+\ell} r^{\sigma+\ell}}{(n-\ell+\sigma)!\,(2\ell+1+\sigma)!}\, e^{-\left(\frac{\gamma}{\ell+1}\right)r}$$

$$\gamma = \alpha(\ell+1) \qquad (3)$$

Recently Weatherford[11] has done an extensive study on the transformation properties of all four types of ETOs and has provided a translation formula for the Hydrogenic Sturmians. The intent of this study is to use the two basis sets, Orthonormal Slaters and Hydrogenic Sturmians, to translate ETOs from one point in space to another and obtain information on the comparative suitability of these sets for multicenter LCAO calculations. This is done by reconstructing an ETO at a field point from the new reference center.

2. RESULTS

The accuracy of the translation formulas for both basis sets: ONS and HS has been checked by reconstructing the orbital at some field point from the new reference center. We have considered the orbtial $n_a=2$, $\ell_a=1$, and $m_a=0$.

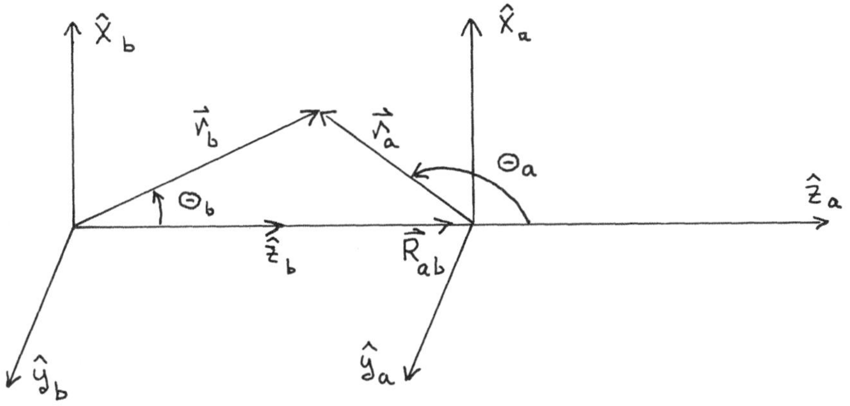

$\phi_a = \phi_b = 0$

$R_{ab} = 0.7a_o$

The values are tabulated in Table-1 and % error has been calculated in each case.

Neither method seems to be capable of microscopic accuracy. These two methods are also checked against the Harris and Michels[11] numerical procedure and it seems that their method is somewhat superior.

The procedure based on Hydrogenic Sturmians is much faster than other procedures since the C's can be calculated and stored once and for all - they are dependent only on the orbital quantum numbers and are independent of the translation distance as well as screening parameters. The C's are on the order of unity for the lower orbitals. The procedure can be easily generalized to the off-axis case.

Table-1.

$R_B(a_o)$ Θ_B	Exact Value	Analytic (ONS) Calculated Value	Analytic (ONS) % Error	Analytic (HS) Calculated Value	Analytic (HS) % Error
R_B=0.30000					
Θ_B = 0°	-0.15128	-0.15401	-1.81	-0.15054	-0.48
= 45°	-0.16169	-0.16645	-2.94	-0.16486	-1.96
= 90°	-0.18441	-0.18530	-0.48	-0.18787	-1.88
= 135°	-0.20173	-0.19999	-0.86	-0.19888	-1.41
= 180°	-0.20755	-0.20723	-0.16	-0.20118	-3.06
R_B=0.62500					
Θ_B = 0°	-0.03926	-0.02866	-27.00	-0.03234	-17.61
= 45°	-0.08727	-0.08559	-1.92	-0.80307	-4.81
= 90°	-0.15452	-0.15498	-0.30	-0.15737	-1.84
= 135°	-0.18936	-0.18911	-0.13	-0.18838	-0.52
= 180°	-0.19870	-0.20169	-1.51	-0.19546	-1.63
R_B=1.25000					
Θ_B = 0°	0.17903	0.17026	4.89	0.15595	12.88
= 45°	0.04206	0.04316	2.62	0.04653	10.62
= 90°	-0.09426	-0.09340	-0.08	0.09574	-1.57
= 135°	-0.14568	-0.14581	-0.09	-0.14494	-0.51
= 180°	-0.15653	-0.15595	-0.37	-0.15844	-1.22
R_B=1.75000					
Θ_B = 0°	0.20730	0.20767	0.18	0.19989	3.57
= 45°	0.07868	0.07893	0.33	0.08201	4.23
= 90°	-0.05997	-0.05976	-0.35	-0.06194	-3.28
= 135°	-0.10971	-0.11000	-0.27	-0.10829	-1.29
= 180°	-0.15653	-0.15595	-0.37	-0.15844	-1.22
R_B=2.25000					
Θ_B = 0°	0.18561	0.18635	0.40	0.18504	0.31
= 45°	0.08116	0.08081	0.43	0.08241	1.54
= 90°	-0.03743	-0.03722	-0.50	0.03869	-3.39
= 135°	-0.09745	-0.07945	-0.00	-0.07824	-1.52
= 180°	-0.08711	-0.08731	-0.23	-0.08976	-3.04
R_B=4.50000					
Θ_B = 0°	0.04796	0.04805	0.18	0.48053	0.19
= 45°	0.02475	0.02470	0.20	0.02468	0.26
= 90°	0.00416	0.00408	-1.72	-0.01482	-2.37
= 135°	-0.01447	-0.01454	-0.46	-0.01481	-2.37
= 180°	-0.01618	-0.01608	-0.62	-0.01445	-10.66

REFERENCES

1. Guseinov, I.I.: 1978, J. Chem. Phys. 69, p.4990.

2. Filter, E.F. and Steinborn, E.O.: 1978, Phys. Rev. A 18, p.1.

3. Jones, H.W. and Weatherford, C.A.: 1978, Int. J. Quantum Chem. Symp. 12, p.483.

4. Antolovic, D. and Delhalle, J.: 1980, Phys. Rev. A 21, p.1815.

5. Guseinov, I.I.: 1980, Phys. Rev. A 22, p.369.

6. Weatherford, C.A. and Jain, B.L.: 1980, Int. J. Quantum Chem. Symp. 14, p.493.

7. Jain, B.L. and Weatherford, C.A.: 1980, Bull. Am. Phys. Soc. 25, p.1142.

8. Filter, E. and Steinborn, E.O.: 1980, J. Math. Phys. 21, p.2725.

9. Slater, J.C.: 1960, Quantum Theory of Atomic Structure I, McGraw-Hill, New York.

10. Rotenberg, M.: 1962, Ann. Phys. N.Y. 19, p.262.

11. Weatherford, C.A.: "Scaled Hydrogenic Sturmians as ETOs", current volume.

AUXILIARY FUNCTIONS FOR STO MOLECULAR INTEGRALS: Si, Ci AND Ei

Frank E. Harris
Department of Physics
University of Utah, Salt Lake City, Utah 84112

The renewed interest in molecular integrals for Slater-type orbitals (STO's)[1] is increasing the urgency of the development of more efficient methods for calculating sine, cosine, and exponential integrals. The sine integral arises in expansions of the Fourier transform of a two-center STO product, and the exponential integral occurs in analytic multi-center integral formulations.[2] The standard analytical methods for evaluating these integrals include the power series expansions for small arguments, and continued-fraction or asymptotic expansions for large arguments.[3] None of these methods are fast for a large intermediate argument range, and the integrals are often evaluated by finite numerically fitted Chebyshev[4] or rational-fraction[5] expansions.

In this communication we wish to point out the possibility of efficient evaluation of these integrals as expansions in squares of spherical Bessel functions. Using x to denote a real variable, the sine integral Si(x) is given by the well-known expansion[6]

$$\mathrm{Si}(x) \equiv \int_o^x \frac{\sin t}{t}\,dt = x\sum_{n=0}^{\infty} [j_n(x/2)]^2 \quad , \tag{1}$$

where j_n is as defined in the Handbook of Mathematical Functions.[7] With the aid of the REDUCE algebraic manipulation computer program,[8] we have found a parallel expression for the cosine integral Ci(x):

$$\mathrm{Ci}(x) \equiv \gamma + \ln x + \int_o^x \frac{\cos t - 1}{t}\,dt = \gamma + \ln x + \sum_{n=1}^{\infty} a_n [j_n(x/2)]^2, \tag{2}$$

where $\gamma \simeq 0.5772156649\ldots$ is Euler's constant and

$$a_n = -(2n+1)\left[\sum_{m=1}^{n} \frac{1}{m} + 1 - (-1)^n\right] \quad . \tag{3}$$

C. A. Weatherford and H. W. Jones (eds.), ETO Multicenter Molecular Integrals, 135–140.

For later convenience we define $a_o = 0$. A proof of Eq. (2) will be given elsewhere.[9]

Viewed as functions of a complex variable z, Si(z) is analytic for all z, but Ci(z) has, due to its logarithmic term, a branch point at z=0; the complex plane is traditionally cut on the positive imaginary axis. The expressions in Eqs. (1) and (2) converge for all z. It is convenient to discuss the exponential integral in terms of the analytic function $E_1(z)$, related to the sine and cosine integrals by

$$E_1(z) \equiv \int_z^\infty \frac{e^{-t}}{t}\, dt = -\mathrm{Ci}(-iz) + i[\mathrm{Si}(-iz) - \frac{\pi}{2}] \quad . \tag{4}$$

This definition places a branch cut for $E_1(z)$ on the negative real axis; the t integration must avoid the negative real axis and proceed toward infinity in the right half-plane. Substituting from Eqs. (1) and (2), and introducing the modified spherical Bessel function $i_n(z) = i^n j_n(iz)$, we have an expansion, convergent for all z:

$$E_1(z) = -\gamma - \ell n\, z + \sum_{n=0}^{\infty} (z-a_n)(-1)^n [i_n(z/2)]^2 \tag{5}$$

For x>0, the exponential integral Ei(x) satisfies

$$\mathrm{Ei}(x) \equiv ⨍_{-\infty}^{x} \frac{e^t}{t}\, dt = -\frac{1}{2}[E_1(-x+i0) + E_1(-x-i0)] \quad , \tag{6}$$

where the integration range is cut at t=0, leading to the "principal value" of the integral. For negative x, $\mathrm{Ei}(x) = -E_1(-x)$. These relationships lead (for both positive and negative x) to

$$\mathrm{Ei}(x) = \gamma + \ell n|x| + \sum_{n=0}^{\infty} (x+a_n)(-1)^n [i_n(x/2)]^2 \tag{7}$$

Equations (1), (2), (5) and (7) converge more rapidly than the corresponding power-series expansions.[10] The necessary j_n or i_n can be obtained by downward recursion starting with the initial approximation i_N or $j_N = 0$.[11] Such a procedure causes the i_n or j_n to be generated with increasing accuracy as n decreases; we have found that N need only be one or two greater than the maximum n values used in the expansions.

The relative convergence rates of various methods for evaluating these integrals are illustrated in Table 1, which lists the numbers of

Table 1. Numbers of terms needed to converge to 9D accuracy for various expansion methods of evaluating Si(x), Ci(x), and Ei(x).

	x	Spher. Bessel	Power Series	Cont. fract.	Asymp.
Si(x)	0.1	2	3	759	e
	0.5	3	5	119	e
	1.	4	6	66	e
	2.	5	7	33	e
	3.	7	9	22	e
	5.	8	12[a]	13	e
	8.	11	16[b]	11	e
	10.	13	20[c]	9	e
	20.	20	35[d]	5	e
	30.	25	--	4	6
	40.	33	--	4	5
Ci(x)	0.1	2	2	513	e
	0.5	3	4	129	e
	1.	4	5	66	e
	2.	6	7	33	e
	3.	7[a]	9[a]	22	e
	5.	9[a]	12[a]	13	e
	8.	11[a]	16[b]	11	e
	10.	13[b]	20[c]	9	e
	20.	21[b]	35[d]	5	e
	30.	28[b]	--	4	6
	40.	34[b]	--	4	5

(Table 1, cont.)

	x	Spher. Bessel	Power Series	Cont. fract.	Asymp.
Ei(x)	0.1	2	5	e	e
	0.5	3	10	e	e
	1.	4	12	e	e
	2.	5	16	e	e
	3.	6	18	e	e
	5.	7	24	e	e
	8.	9	32	e	e
	10.	11	35	e	e
	20.	15	53	e	e
	30.	18	67	5	12
	40.	21	83	3	10
	-0.1	2	6	284	e
	-0.5	4	10	71	e
	-1.	5[a]	13[a]	32	e
	-2.	6[b]	17[b]	20	e
	-3.	7[b]	21[b]	13	e
	-5.	10[c]	29[c]	9	e
	-8.	15[d]	41[d]	7	e
	-10.	-	-	7	e
	-20.	-	-	5	e
	-30.	-	-	4	12
	-40.	-	-	3	8

a. 1D lost in differencing
b. 2D lost in differencing
c. 3D lost in differencing
d. 5D lost in differencing
e. Does not yield required accuracy

terms needed to generate results of 27-bit accuracy (full-precision floating-point numbers in a typical 36-bit machine). The continued-fraction method is that reported by Stegun and Zucker,[3] but modified to avoid scaling difficulties. The asymptotic and power-series methods are based on the usual formulas.[10] We note that both the Bessel-function and power-series expansions are subject to differencing errors in certain argument ranges. For Si and Ci the Bessel expansion is superior in this regard to the power series because the former consists of terms all of the same sign, while the latter can possess terms orders of magnitude larger than their sum. These differencing errors can be ameliorated by performing intermediate calculational steps at double precision. Bearing in mind the above, plus the fact that (at equal precision) each added term in the Bessel-function expansion is comparable in computation effort with about two terms in the asymptotic or power series or one to two continued-fraction steps, we conclude that for Si at 9D the simplest good approach will be to use the Bessel-expansion and continued-fraction methods, with the optimum cross-over point between methods dependent on the machine characteristics and the details of the computer coding. For Ci, there is an inherent differencing problem between $\gamma + \ln x$ and either the Bessel or power series; this fact may make optimum the use of the power series at double precision for $|x|$ smaller than the range for which the continued-fraction method is efficient.

For Ei(x) with $x>0$, the Bessel expansion will be optimum by a considerable margin everywhere except for x values large enough that the continued-fraction method becomes acceptable. For negative x, the best methods will be the Bessel expansion for small $|x|$, and the continued-fraction method for intermediate and large $|x|$. Differencing errors in the Bessel expansion will have to be considered in determining the optimum cross-over point and in deciding whether intermediate steps should be carried out in double precision. For almost the entire range of x values, the methods here advocated for Ei(x) differ from those ordinarily employed.

Computer programs based on the foregoing analysis have been deposited in the Quantum Chemistry Program Exchange.[12]

Acknowledgement

The author wishes to acknowledge illuminating discussions with Dr. N.H.F. Beebe. Computations were carried out on the DEC 2060 at the College of Science Computer of the University of Utah.

REFERENCES

1. Weatherford, C.A. and Jones, H.W. (Eds.): Proceedings of the First International Conference on ETO Multicenter Molecular Integrals, D. Reidel, Holland (present volume).

2. Kay, K.G. and Silverstone, H.J.: 1969, J. Chem. Phys. 51, p.956, 4287; 1970, J. Chem. Phys. 53, p.4269.

3. Stegun, I. and Zucker, R.: 1976, J. Research Nat. Bur. Stand. (U.S.) 80 B, p.291.

4. Bulirsch, R.: 1967, Numer. Math. 9, p.380.

5. Hastings, C.: 1955, Approximations for Digital Computers, Princeton Univ. Press, Princeton, N.J..

6. Abramowitz, M. and Stegun, I.A.: 1964, Handbook of Mathematical Functions, Appl. Math. Ser. 55, Nat. Bur. Stand. (U.S.), Formula 10.1.52.

7. Reference 6, Chapter 10.

8. Hearn, A.C.: 1973, REDUCE-2 User's Manual, Department of Computer Science, Univ. of Utah.

9. Harris, F.E.: to be published.

10. Reference 6, Chapter 5.

11. Carbato, F.J.: 1956, J. Chem. Phys. 24, p.452.

12. Quantum Chemistry Program Exchange, Indiana University, Bloomington, Ind. 47405 (Submitted).

ON AUXILIARY FUNCTIONS IN MOLECULAR INTEGRALS

Milan Randić
Ames Laboratory, US-DOE, Iowa State University, Ames, Iowa 50011, and Department of Mathematics, Drake University, Des Moines, Iowa 50311

A brief review of work on an analytical solution to three center nuclear attraction integrals and relationship among selected auxiliary functions is presented. The review is preceded by an outline of a few concepts from graph theory -- the subject of current intensive interest of the author. Graphs already play useful role in molecular calculations, even in some problems involving molecular integrals and graph theory deserves a better exposure. In the spirit of graph theoretical tradition, the problem of solving of four center integrals has been here exalted to a conjecture that such integrals can be evaluated analytically. While believers will try to prove the conjecture, now there is also a burden on non-believers to disprove it! In a more solemn direction, the contribution suggests that overlap integrals play a role of auxiliary functions and some hope is expressed that additional molecular integrals will be expressed in terms of general overlap integrals or as some function of overlap integrals.

1. INTRODUCTION

Molecular integrals have a long history. Already the first papers on quantum chemistry indicated that accompanying molecular integrals might be difficult. Hence the efforts to solve an simplify molecular computations comprised to great extent the efforts to solve molecular integrals. It has been observed at this Conference that the chemical community at large appears to be interested in exploiting the results of these continuous efforts, yet it also appears unwilling to grant support for such endeavors[1]. This is to be deplored and the first part of this exposition is aimed to draw attention of the general chemist to our problems. The situation is particularly undeserving, since about 1970 when much of the activity on Slater-type integrals was still flourishing, a surrogate scheme using Gaussian-type functions emerged and was widely accepted. This was less due to merits of Gaussian functions and more due to the prevailing impatience of consumers who use and misuse theoretical chemistry. Not everyone ought to be enthusiastic about the challenge of unsolved problems, such as multi-center ETO molecular integrals. But everyone should be aware

C. A. Weatherford and H. W. Jones (eds.), ETO Multicenter Molecular Integrals, 141–155.

that (1) No one has established that molecular integrals based on STO can be solved in a practical manner; or that such solutions cannot be competitive to the alternative approach using Gaussian type functions; and (2) Claims as to the adequacy of GTO have never been critically examined. Gaussian functions are not the first choice in theoretical chemistry. They are used -- and this ought to be emphasized from time to time to those who use such functions, as a tool without much understanding of the computations involved -- primarily because molecular integrals can be evaluated, not because they possess desirable properties. Today this may be a valid reason for their use, but tomorrow they may be thought of as bastard surrogates, which served their purpose in the transition period, have no longer viable merits and will fall into oblivion. Another extreme situation that tomorrow may bring is development of conclusive evidence of the intractable character of difficult molecular integrals using STO, evidence based on proof not opinions. Such proof may be even more difficult to obtain than an attempted solving of integrals, but it is as essential as the evaluation of an integral would be to resolve the dilemma on the future of STO.

This Conference is evidence that today there is still continuing interest in STO molecular integrals. We heard opinions that even with the present unresolved difficulties of some many center integrals (and use of approximate numerical integration schemes) use of STO is competitive to GTO, or at least not so unfavorable as one has been lead to believe. No doubt in the future many comparisons will be made between the two methodologies, as well as between various approaches to selected molecular integrals. In this contribution we will illustrate one such particular approach centered around three-center nuclear attraction integrals and alternative auxiliary functions, which is the main theme of this presentation. Particularly, we will see that overlap integrals can be viewed as auxiliary functions, which, if shown more general, will further justify revised interest and novel approaches to overlap integrals, to which in recent years Professor H.W. Jones, who organized this Conference, has significantly contributed. Before going into this more technical side of the paper, a brief outline of the Graph Theoretical approach to chemical problems, which may well in the future, find its way into complex molecular calculations, will be presented. Clearly the intent is to advertize Graph Theory and perhaps illustrate some of its flavor, raise some expectations. But how useful and practical such directions will be mostly depend on those who are sufficiently enticed to try them in problems of their interest.

2. GRAPH THEORY AS A TOOL

Informally graphs are objects defined by a set of elements and a binary relation[2]. Pictorially they can be represented as points (customarily shown as small circles) and connecting lines. Hence, superficially they resemble molecular valence diagrams, if atoms are viewed as vertices (points) and bonds as lines (edges) signifying binary relations. Graph Theory as a mathematical discipline is concerned with properties of such objects and with various manipulations of such graphs. Several common manipulations of interest in chemistry have been listed in Table 1. Observe that some of the shown operations (on graphs) may prerequisit some other operations. Comparing of graphs for example needs prior characterization. The operation may appear trivial, everyone in his work makes comparisons. Here, the operations are rigorously defined, conditions are sought under which comparisons are possible, meaningful, or not possible. Comparison of two numbers is trivial, but comparison of two structures, two sequences, and two general mathematical objects is far from trivial. It is common to find two objects that cannot be meaningfully compared, but not everyone is aware that even under these circumstances, both objects may be comparable to a third object. Table 1 is far from complete, an additional listing of operations can be found elsewhere[3]. Perhaps any such listing of operations will be deficient. It has been suggested at this Conference that Planning should be included[4]. Definitely so. Flowcharts used in computer programming are for example illustrations of planning. They have been a legitimate object of mathematical analysis which involves elements of graph theory. Although it has a long history and considerable potential, graph theory as a tool has not been widely used in chemistry. This is likely to change. Most people in computational chemistry have already seen the use of graphs in the work of Shavitt[5] and others[6]. Incidentially, those graphs have some special properties and should more precisely be referred to as lattices. This is not because of their regular shape resembling a physical lattice, but because they represent partial order (with a single dominant configuration and a single configuration dominated by all others). Shavitt's graph or lattice is however mainly a notational device which facilitates computation of associated matrix elements. However graphs can be used operationally, as is illustrated with graphs depicting Wigner 3nj symbols, introduced in computational chemistry already in 1957 by Levinson. (7)

In quite general terms one may say that problems involving considerable complexity of combinatorial origin will be suited for graph theoretical analysis. Large simplifications can then follow. This is analogous to use of group theory in problems of high symmetry. If a problem has limited symmetry one can probably resolve it using some symmetry notions intuitively, without the powerful tool of the group theory, and perhaps even without recognizing group theoretical arguments. Similarly, if a problem has limited combinatorial content it can be resolved without an explicit use or mention of the graph theory, the power of which becomes evident as the complexity of the problem increases. We hope here only to show some of the "flavor" of graph theory.

Novel insights frequently follow because of one's novel point of view. Graph theory offers such novel viewing of old problems. It is not uncommon in applications of graph theory to recognize a relation that has been overlooked for too long. As an illustration, take the

TABLE 1. A Selection of Operations on Graphs of Interest in Chemistry

AUTOMORPHISM	Finding the symmetry of a graph.
CHARACTERIZATION	Assigning a set of parameters (or a single index) reflecting some structural features to a graph.
CLASSIFICATION	Grouping structures according to preselected features.
COMPARISON	Watching for difference in selected graph invariants.
COMPOSITION	Combining constitutents (such as fragments) into a structure.
CONSTRUCTION	Building of graphs of prescribed properties in a systematic way.
DISCRIMINATION	Differentiation of graphs on the basis of a singular property.
DISSECTION	Dismembering of a graph into preselected elementary parts.
ENCODING	Assigning to a graph a name (usually derived by counting some graph invariants).
ENUMERATION	Counting graphs of particular size and structure.
FRAGMENT SEARCH	Identifying and counting the occurance of a fragment in a larger structure (subgraph isomorphism).
ISOMORPHISM	Verifying the connectivity of two graphs in order to establish if they are identical.
ORDERING	Sequencing structures according to selected dominant property.
PARTIAL ORDER	Relating structures in a hierachical order.
RECONSTRUCTION	Recovery of a graph from fragmentary data.
SIMILARITY	Appraisal of the degree of overlap.

well known HMO calculations, which have received heavy attention during the past 50 years. One does not expect novelties in such a well-trod field, yet relatively lately (in 1973)[8] it was recognized that there are molecules, such as 1,4-divynilbenzene and 2-phenylbutadiene:

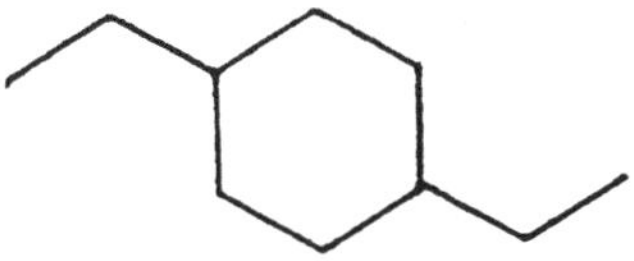

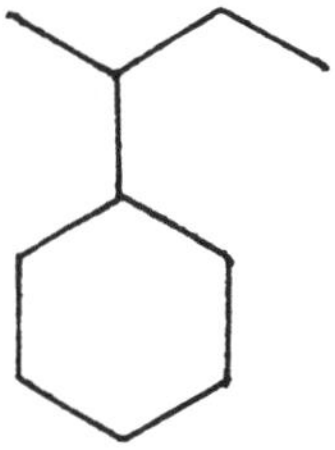

which have all molecular orbital energies the same. Many chemists are even today not aware of this fact, which could have been discovered long ago. For example the widely used "Dictionary of Pi-Electron Calculations" of Coulson and Streitwieser[9] (with extensive help from Poole and Brauman) lists the two molecules mentioned only a few pages apart! It is irrelevant that HMO method has since been found of limited use, because the same mathematical apparatus is of interest in other problems of chemistry and chemical physics. In graph theory we are intrigued by such coincidences and are interested in reasons why the above anomaly (or regularity?) occurs. Answering such questions is of interest even in the field of chemistry of conjugated hydrocarbons, despite narrowness of HMO when discussing details of bonding in such molecules. One may reason that understanding all facets of simple schemes is a prerequisite for understanding of more advanced models. At least a comparison could then show which part of the results can be truly attributed to topological origin and which go beyond the concept of molecular connectivity. This would be analogous to a comparison between Hartree-Fock calculations which indicate the limit of an independent particle model and calculations which go beyond Hartree-Fock approximation. Just as the Hartree-Fock approximation represents a well defined and useful step in solving Schrödinger equation featuring an important reference point so the Hückel model (the approximation of the nearest neighbor interaction used by Hückel is due to Felix Bloch[10]) plays a similar role with respect to topological factors featuring the reference point delineating topological and geometrical effects. Hence, the suggestion that the concept of isospectral graphs is not relevant in chemistry[11] is at best misleading. But in addition the study of graph spectra is of interest in other areas of physics and chemistry.

Another closely related question is that of the occurrence of common eigenvalues in different graphs. Why, for example, do the two structures shown below have numerous common eigenvalues? This problem received some attention, some special cases have been considered,[12] additional results await publication.[13]

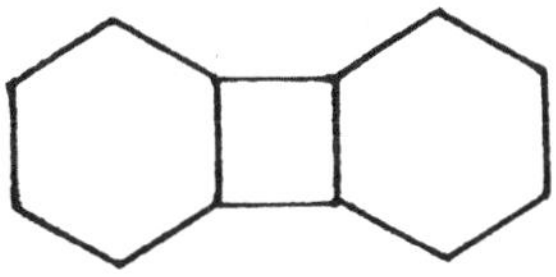

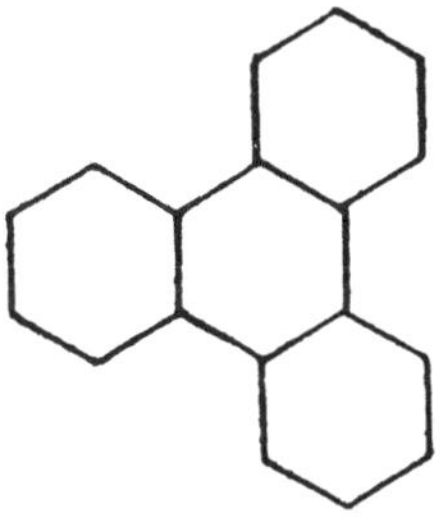

Graph Theory is one of the few mathematical disciplines abundant in unsolved problems. Consider the skeleton[14]

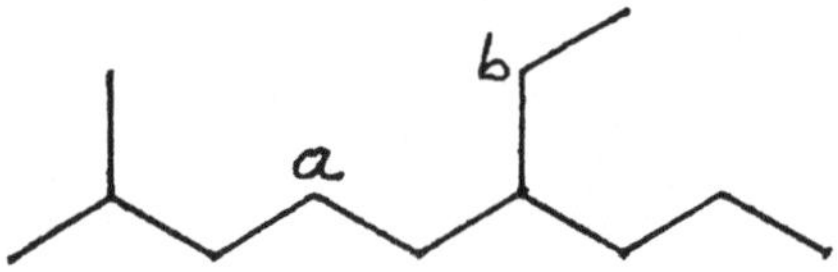

If one counts self-returning random walks (i.e., random walks which start and end on the same atom) for such a network, surprisingly one finds two non-equivalent sites a and b which show the same number of walks of any length. Why? Even more intriguing is the case of the two graphs shown below[15]:

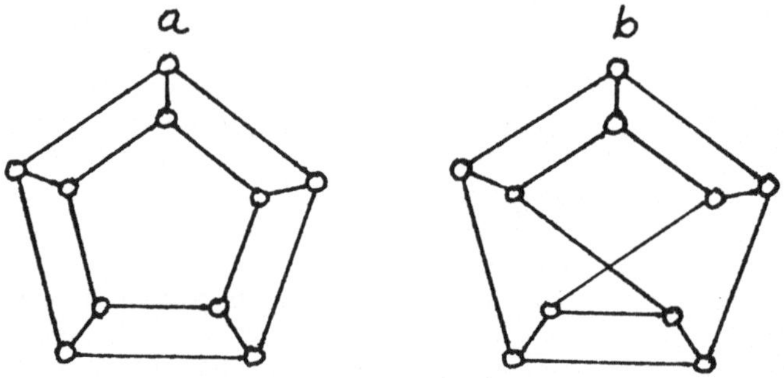

They show the same count of self-returning walks for the marked vertices, at least this has been verified for walks of length 20 and less. Observe that here the recursive formulas for evaluation of higher walks (i.e., the characteristic polynomials as used in Cayley-Hamilton theorem) are different, since the graphs are not isospectral, nevertheless mysteriously for the indicated points the random walks persist to be same[16].

Let's end this brief excursion to graph theory by mentioning a few famous problems in graph theory. Traditionally in mathematics famous problems are those which can be simply formulated so that even a non-mathematician can fully understand the problem, but the solution of which has so far eluded even most gifted mathematicians. With the recognition of Gödel's Theorem, opinions have been heard that perhaps some such problems may have solution (i.e., the conjecture can be true or false), but still may be beyond the domain of provable theorems. Among the most famous such problems in graph theory are the following:
(1) The Four Color Conjecture -- shortly 4CC -- which states that any (goegraphical) map regardless of the number of states or shapes of boundaries can be colored in four colors[17].
(2) Hamiltonian Circuit Problem -- known in commercially oriented societies as the problem of the traveling saleman -- requires characterization of networks for which one can find closed path in which all

sites (cities) have been visited once[18].
(3) Ulam's Reconstruction Conjecture -- requires proof that a collection of qualified subgraphs (derived by erasing each time a single vertex and all its incident edges) suffices for full recovery of the graph.

All the three problems are known to have fascinating qualities and are capable of trapping experienced as well as unexperienced in a prolonged futile struggle. The graph reconstruction problem was first suggested as an exercise for students in a mathematical text[19]. It has since become famous, because no student and no professor has yet solved it! Perhaps difficult problems ought to be given as an exercise to students -- in order to attract proper attention. How many chemists and mathematicians are aware of the difficult problem of four center integrals over STO? Yet the four color problem is discussed even in barber shops!

Let's consider the graph reconstruction problem a bit more. In Figure 1 we illustrate Ulam's subgraph for a graph

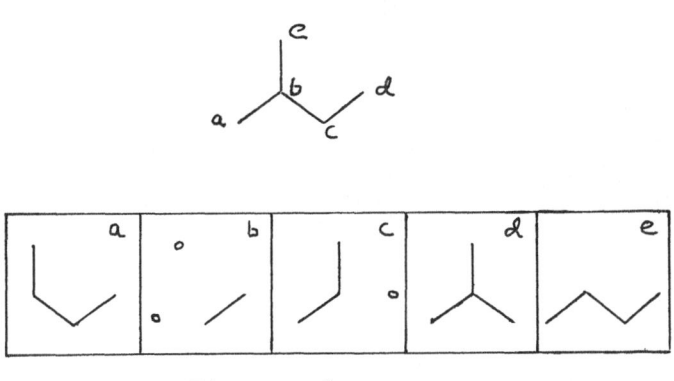

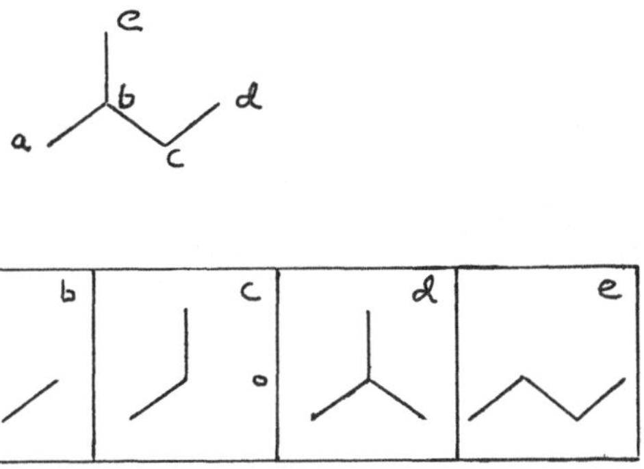

Figure 1

We have labeled the vertices and have kept the forms of subgraphs deliberately preserving the correspondence with the graph shown above in order to facilitate comparison. Generally the subgraphs will be given in an arbitrary from with no apparent relationship between different subgraphs. That the reconstruction is a difficult problem has been generally accepted. Here the label "difficult" has to be interpreted in the spirit of Euler's comment on the problem of Königsberg bridges[20]. The problem is difficult if the amount of work increases as n! with the size of the problem. An exhaustive examination of all combinatorial possibilities is, however, not considered as an acceptable approach. Some such difficult problems have been solved, others wait for a solution. In the case of graph reconstruction (which has been solved only for some special classes of graphs) one may take a pessimist's point of view[21]: "The reader is urged not to try to settle the conjecture, because it is rather difficult", or an optimist's point of view[22]: "A clever amateur has equal chance to solve the problem as experienced professionals." To these two extremes we can add yet more advice. After this author's seminar on graph theory in chemistry at Chapel Hill at which Ulam's conjecture was mentioned Professor Parr gave a warning to his students: "I forbid you to try to solve this problem." In view of the "contagious" character of the graph reconstruction "disease" such an advice appears to be a prudent step whenever there are large groups of people in close contact whose existence depends on visible accomplishments.

3. CONJECTURE ON FOUR CENTER INTEGRALS

The unsolved problem of quantum chemistry, that of the evaluation of four center molecular integrals involving STO can be compared to the famous problems of graph theory. Much of what has been said about the famous problems of graph theory could have been said also about our unsolved problem of four center integrals. Thus one can urge chemists not to solve it, or one can say that clever amateurs and experienced professionals have the same chance to solve the problem. Finally those whose support and existence depend on visible results should at best postpone their desire to tackle such problems. The problem is easy to formulate but difficult to solve. Hence the situation that so closely parallel those of the famous problems in mathematics deserves similar glorification! In the spirit of graph theory we are therefore inclined to humbly suggest that the long standing unsolved problem of four center integrals be elevated to the status of a conjecture:

The Four Center Conjecture -- "Two electron four center STO molecular integrals can be solved analytically."

Besides the advantage of perhaps becoming a subject of a wider recognition, the conjecture now calls for attempts not only of being proved but equally of being disproved. Hence, some burden is also on those who are skeptical, to justify their criticisms. This elevation of the four center integral problem to a conjecture should have been made perhaps 50 years ago, and if it were also included as a student excercise in a text, perhaps it would have been already solved, or at least became famous! Despite these past political omissions it seems that 4CC (four center conjecture, not four color conjecture!) deserves to be considered as one of the famous problems in chemistry. The problem can be understood by everyone, but a solution has so far been evasive. Nobody proved that the problem cannot be solved -- hence an optimism is in order that one day we may find the solution.

4. ANALYTICAL EVALUATION OF THREE-CENTER NUCLEAR ATTRACTION INTEGRALS

In this and the next section we will review some results in the area of molecular integrals to which this author contributed. Although major results and more details are available[23,24] the decline in activities in ETO integrals after about 1970 caused these available results not to be sufficiently known. For example Guseniev's work[25] in 1975 on the analytical evaluation of three center nuclear attraction integrals does not mention work of Bosanac and Randic,[23] which we believe is not intentional. We will not discuss here the advantages and merits of alternative schemes, but will rather point to the particular individuality of one such approach which unexpectedly produced relatively simple expressions for the integrals.

First, let me indicate how our interest in Zagreb originated in molecular integrals. We had been aware of limited computational resources at the time in Zagreb. The institute "Rudjer Bosković" just acquired its first computer (French made SDS with 16 K memory capacity) and it was clear that we had to have similarly limited objectives. Because of previous experience with the single center method derived in collaboration with D.M. Bishop,[26] extending such methodology to a "two-center method", with two heavy atoms appeared to be a potentially interesting problem. Such a method would then be useful for close and rigorous examination of molecules like ethane, ethylene, acetylene, diborane, methyl alcohol, hydrogen peroxide, disilane, and many other hydrides -- a significant amount of chemistry. At the time two electron two center integrals were available (e.g., Roothaan[27] and Reudenberg[28] works). Some examination of three center integrals appeared desirable. We will outline here the principal steps of the particular evaluation that we developed. The late professor C.A. Coulson, who has been involved in this kind of integrals for quite a while, acknowledged our approach with the comment that he believes our approach to be the most simple and most elegant such approach to three center nulcear attraction integrals, the area as he said,[29] in which he has been engaged for over 35 years. This is, of course, just one opinion, and may not even be quite correct, but it was nice to hear. As will be seen, at one point in the rather standard procedure we decided to try substitution of variables which resulted in the recovery of the limits of integration of certain integrals. The integrals then corresponded to standard functions A_n and B_m of overlap integrals, the only novelty was that the n index could also take negative values. However the corresponding exponential integrals are not difficult than standard A_n auxiliary functions.

The integral we consider is:

$$I = \int \phi_A r_C^{-1} \phi_B d\tau$$

with ϕ being Slater-type orbitals:

$$\phi = N\, r^{n-1} \exp(-\zeta r)\; Y_{\ell m}(\Theta,\phi)$$

with the usual meaning for all symbols. We start with a standard procedure of introducing elliptic coordinates

$$\xi = \frac{r_a + r_b}{R}; \quad \eta = \frac{r_a - r_b}{R}; \quad \phi$$

and use Neumann's expansion of $1/r_c$

$$r_c^{-1} = \frac{4}{R} \sum_{o}^{\infty} \sum_{-\lambda}^{+\lambda} (-1)^n \frac{(\lambda-n)!}{(\lambda+n)!} P_\lambda^n(\xi_>) \, Q_\lambda^n(\xi_>) \, .$$

$$P_\lambda^n(\eta) \, P_\lambda^n(\eta_c) \exp \, i(\phi-\phi_c)n$$

(for more details consult the indicated paper of Bosanac and Randić). The above expansion leads then to the expression

$$I^1 = \frac{4}{R} K_{AB} \sum_{M}^{\infty} \sum_{o}^{N-M} \sum_{o}^{N-M} (-1)^n \frac{(\lambda-M)!}{(\lambda+M)!} a_{nj} \, P_\lambda^M(\eta_c) \, .$$

$$\int_{-1}^{+1} P_\lambda^M(\eta) \exp(-pt\eta)\eta^j (1-\eta^2)^{M/2} d\eta$$

$$\Big\{ Q_\lambda^M(\xi_c) \int_1^{\xi_c} P_\lambda^M(\xi)\xi^n \exp(-p\xi) (\xi^2-1)^{M/2} d\xi \; +$$

$$+ P_\lambda^M(\xi_c) \int_{\xi_c}^{\infty} Q^M(\xi)\xi^n \exp(-p\xi) (\xi^2-1)^{M/2} d\xi \Big\}$$

with a_{nj} and K_{AB} being constants. The above integral is essentially a combination of integrals I(P'), I(P) and I(Q) of the form: $I(P') \, [I(P) + I(Q)]$; where

$$I(P) = \int_1^{\xi_c} P_\lambda^M(\xi)\xi^n \exp(-p\xi) (\xi^2-1)^{M/2} d\xi$$

$$I(Q) = \int_{\xi_c}^{\infty} Q_\lambda^M(\xi)\xi^n \exp(-p\xi) (\xi^2-1)^{M/2} d\xi$$

$$I(P') = \int_{-1}^{+1} P_\lambda^M(\eta)\eta^j \exp(-pt\eta) (1-\eta^2)^{M/2} d\eta$$

At this point in the analysis we depart from other authors. Instead of considering variable integration limits we will try to transform the integrals to forms with fixed integral limits. This can be made using the following two steps:

(1) $\int_1^{\xi_c} F(\xi) d\xi = \int_1^{\infty} F(\xi) d\xi - \int_{\xi_c}^{\infty} F(\xi) d\xi$

(2) Substitute $\xi\xi_c \to \xi$ which gives

$$\int_{\xi_c}^{\infty} F(\xi)d\xi = \xi_c \int_1^{\infty} F(\xi\xi_c)d\xi$$

The first step converts the initial integral to integrals having the upper limit infinity. The second step restores the lower limit of the second integral to 1, making it independent of the position of the center C. The variable parameter is absorbed with other geometrical parameters defining p and pt. This latter step, the substitution, is the critical step leading to very simple auxiliary functions A_n and B_m. The substitution works here because of the properties of Slater-type functions and because the other contributing parts can be expanded as a power series. Therefore the character of the integrand remained the same as it occurs in generalized overlap integrals.

The remaining part of the analysis is straight forward. Let's consider one of the three integrals only:

$$I(P) = \int_1^{\infty} P_\lambda(\xi)\xi^n \exp(-p\xi)(\xi^2-1)^{M/2}d\xi -$$

$$-\xi_c^{n+1} \int_1^{\infty} P_\lambda^M(\xi\xi_c)\xi^n \exp(-p\xi_c\xi)(\xi^2\xi_c^2-1)^{M/2}d\xi$$

Observe that polynomial in $(\xi\xi_c)$ differs from polynomial in ξ only by coefficients. Similarly expansion of $(\xi^2-1)^{M/2}$ and $(\xi^2\xi_c^2-1)^{M/2}$ will differ for corresponding terms only in coefficients. The exponential functions $\exp(-p\xi_c\xi)$ and $\exp(-p\xi)$ can be viewed as functions which differ in an effective screening parameter p. Hence, after substituting we obtain

$$I(P) = \Sigma\,\Sigma\,(\text{coeff})_m A_m(P) -$$

$$- \Sigma\,\Sigma\,(\text{coeff})_m \xi_c^{m+1} A_m(P\xi_c)$$

The auxiliary functions derived are well known $A_m(p)$ and $A_m(p')$. The integral I(Q) can be approached in precisely the same way again leading to the same auxiliary functions, while the integral I(P') leads, in the usual way, to auxiliary integrals $B_m(pt)$.

Remarkably we thus arrive at an analytical expression for the three-center nuclear attraction integrals using only very familiar auxiliary functions. Remembering now the long history of calculations of three-center nuclear attraction integrals, we see that it may take time, many years, before a relatively simple scheme evolves. These integrals have been termed "difficult", but have been found relatively simple. In the same spirit one can anticipate that "difficult" four center two electron integrals may eventually be solved, moreover after some time even an alternative "simple" route for their evaluation may emerge.

5. ON AUXILIARY FUNCTIONS

Clearly the terms "difficult", "simple", and "elegant" are all to a considerable extent subjective, and are used here to emphasize that three-center nuclear attraction integrals have been expressed in terms of auxiliary functions previously known from the study of simpler overlap integrals. Reducing new to something already known is in accord with Plato's position that cognition is mainly recognition. We have recognized the A_n and B_m functions in a somewhat more difficult problem. One can take the even more generalized position and consider the problem of a particular molecular integral resolved if it can be reduced to combination of more familiar integrals. Hence the effort to solve three-center two electron integrals may be directed to relating these integrals to three-center one electron integrals. More specifically we will outline now how overlap integrals can play the role of auxiliary functions.

We start with the Coulson and Barnett's Z function, which is defined as:

$$Z_{m,n,\ell+\frac{1}{2}}(K,\tau) = \int_0^\infty e^{-Kt}\, \xi_{m,n}(t,\tau)\, t^{\ell+\frac{1}{2}} dt$$

ξ arise from the expansion of ϕ_b about the other center. Here t is the scaled distance ($t=K_b r_a$) and τ is fixed distance ($\tau=K_b R_{ab}$). The $\xi_{m,n}$ are closely related to Löwdin's α-function.

On examining the work of Coulson and Barnett, one finds that $Z_{m,n,\ell+\frac{1}{2}}$ also appears in evaluation of overlap integrals, although overlap integrals do not present unusual problems and the "machinery" of Z function has been introduced because of more difficult integrals. In Table 2 we have listed selected overlap integrals expressed in terms of Z functions. Observe that in all cases a simple proportionality holds. Moreover, one of the orbital is always an s-function. The simple proportionality allows an alternative way of viewing Table 2 -- one can interpret Z functions as selected overlap integrals. This then allows:

(1) expressing integrals already given by Z functions as combination of overlap integrals; and

(2) examining novel relations among integrals, including also novel relations among overlap integrals.

There have been reports concerning both counts. Harris expressed two-center Coulomb integrals via overlap integrals, and in a recent publication of an earlier work Harris and Randić extended the same approach to hybrid integrals. Of considerable interest is the problem of expressing three-center nuclear attraction integrals via overlap

TABLE 2. A selection of $Z_{m,n,\ell+\frac{1}{2}}$ functions expressed as overlap integrals.

$Z_{0,0,\frac{1}{2}} = \frac{1}{2}\,(\tau/K)^{\frac{1}{2}}\,(0s,0s)$	$Z_{1,0,\frac{3}{2}} = \frac{1}{4}\,(\tau/K^3)^{\frac{1}{2}}(1s,1s)$
$Z_{1,0,\frac{1}{2}} = \frac{1}{2}\,(\tau/K)^{\frac{1}{2}}\,(0s,1s)$	$Z_{1,0,\frac{5}{2}} = \frac{1}{4}\,(3\tau/K^5)^{\frac{1}{2}}\,(2s,1s)$
$Z_{0,1,\frac{1}{2}} = \frac{1}{2}\,(\tau/3K)^{\frac{1}{2}}\,(0p,0s)$	$Z_{1,1,\frac{5}{2}} = \frac{1}{4}\,(\tau/K^5)^{\frac{1}{2}}\,(2p,1s)$
$Z_{1,1,\frac{1}{2}} = \frac{1}{2}\,(\tau/6K)^{\frac{1}{2}}\,(0p,1s)$	$Z_{2,0,\frac{5}{2}} = \frac{3}{4}\,(\tau/K^5)^{\frac{1}{2}}\,(2s,2s)$
$Z_{0,1,\frac{3}{2}} = \frac{1}{2}\,(\tau/6K^3)^{\frac{1}{2}}\,(1p,\,0s)$	$Z_{2,1,\frac{5}{2}} = \frac{1}{4}\,(3\tau/K^5)^{\frac{1}{2}}\,(2p,2s)$
$Z_{1,1,\frac{3}{2}} = \frac{1}{4}\,(\tau/3K^3)^{\frac{1}{2}}\,(1p,1s)$	$Z_{1,2,\frac{7}{2}} = \frac{3}{8}\,(2\tau/K^7)^{\frac{1}{2}}\,(3d,1s)$

TABLE 3. Some relations between overlap integrals.

$$(2p,2p)_\sigma = \tau(2p,1s) - (1/K)\left[(5/2)^{\frac{1}{2}}(3s,1s) + 2^{\frac{1}{2}}\,(3d,1s)\right]$$

$$(2p,2p)_\pi = (1/K)\left[(5/2)^{\frac{1}{2}}(3s,1s) - 2^{-\frac{1}{2}}\,(3d,1s)\right]$$

$$(3s,2p)_\sigma = \tau(3s,1s) - (14^{\frac{1}{2}}/3K)(4p,1s)$$

$$(3p,2p)_\pi = (14/3)^{\frac{1}{2}}\,(1/K)\left[(4s,1s) - (1/5)(4d,1s)\right]$$

integrals. The significance of the second topic: search for novel relations among integrals, is mainly in reducing the number of independent integrals needed to be evaluated. In Table 3 we illustrate a few such relations for overlap integrals. Such relations have been overlooked in the past, when they could be valuable for checking numerical magnitudes of different integrals. Relations between overlap integrals shown in Table 3 follow from relations among electron charge distributions on two center.

We would like to conclude this expository contribution by expressing optimism in the future role of overlap integrals as auxiliary functions. In a way the overlap integrals appear very natural auxiliary functions for molecular calculations. If support for research on integrals is restored and this kind of problems appreciated by chemists I anticipate another international conference on ETO molecular integrals in the not so distant future. The title of my future contribution I could announce now: "Overlap integrals as auxiliary functions in evaluation of many center ETO molecular integrals." Our optimism may or may not be justified but the real problems are not "difficult many center integrals". The real problem is the lack of appreciation of the subject by the chemical community in general, particularly a hypocrisy among those who accept various quantum chemistry computer "packages", and use them in their areas of interest, but have forgotten the old Chinese proverb: "When you drink water, remember the man who dug the well". This is indeed disappointing, particularly because the present computational "well" appears to be somewhat shallow and there are people around willing to dig and experienced in this kind of work! But support is lacking.

REFERENCES

1. Comment made by several participants.

2. Wilson, R.J.: 1972, Introduction to Graph Theory, Academic Press, New York.

3. Randic, M.: 1979, MATCH 7, p.3.

4. Comment made by Professor Roothaan at the Conference.

5. Shavitt, I.: 1977, Int. J. Quant. Chem. Symp. 11, p.131.

6. Duch, W. and Karwowski: 1981, Lecture Notes in Chemistry #22, p.260.

7. Yutsis, A.P., Levinson, I.R. and Vanagas, V.V.: 1962, Mathematical Apparatus of the Theory of Angular Moments, Israel Program for Sci. Translations, Jerusalem. In fact, already in 1954 Ord-Smith depicted 3nj relations as graphs but did not consider possible manipulations with graphs; Ord-Smith, R.J.: 1954, Phys. Rev. 94, p.1227.

8. Zivkovic, T.: 1973, A report at Theoretical Chemistry Summer School in Leningrad; see: Zivkovic, T., Trinajstic, N. and Randic, M.: 1975, Mol. Phys. 30, p.517.

9. Coulson, C.A. and Streitwieser, A.: Dictionary of π-Calculations, W.H. Freeman and Co., San Francisco.

10. Bloch, F.: 1930, Zeit. f. Phys. 61, p.206.

11. Heilbronner, E. and Jones, T.B.: 1978, J. Amer. Chem. Soc. 100, p.6506.

12. D'Amato, S.S., Gimarc, B.M. and Trinajstic, N.: (in press), Croat. Chem. Acta.

13. Randic, M.: unpublished results.

14. Randic, M.: 1980, J. Comput. Chem. 1, p.386.

15. Randic, M. and Graovac, A.: unpublished result.

16. Fink, A.: (Iowa State University, Ames) advanced some algebraic theorems that may account for the situation.

17. Biggs, N.L., Lloyd, E.K. and Wilson, R.J.: 1976, Graph Theory 1736-1936, Clarendon Press, Oxford; Chapter 6 and 9.

18. Ref. 17, chapter 2.

19. For a review see: Bondy, J.A. and Hemminger, R.L.: 1977, J. Graph Theory 1, p.227.

20. Ref. 17, chapter 1.

21. Harary, F.: 1969, Graph Theory, Addison-Wesley, Reading, Mass,; footnote on p.12.

22. O'Niel, P.V.: 1970, Amer. Math. Monthly 77, p.35.

23. Bosanac, S. and Randic, M.: 1972, J. Chem. Phys. 56, p.337.

24. Randic, M.: 1972, J. Chem. Phys 56, p.5858.

25. Guseinov, I.I.: 1976, J. Chem. Phys. 65, p.4718.

26. Bishop, D.M. and Randic, M.: 1966, J. Chem. Phys. 44, p.2480; 1966, Mol. Phys. 10, p.517.

27. Roothaan, C.C.J.: 1951, J. Chem. Phys. 19, p.1445.

28. Ruedenberg, K.: 1951, J. Chem. Phys. 19, p.1495.

29. Private correspondence.

30. Barnett, M.P. and Coulson, C.A.: 1951, Phil. Trans. Roy. Soc. (London) 243, p.221.

31. Löwdin, P.O.: 1956, Adv. Phys. 5, p.1.

32. Harris, F.E.: 1969, J. Chem. Phys. 51, p.4770.

33. Randic, M. and Harris, F.E.: 1979, Croat, Chem. Acta 52, p.347.

DISTRIBUTION OF MATRIX ELEMENTS FOR WAVEFUNCTION COMPUTATIONS OF LARGE MOLECULAR SYSTEMS

Peter Habitz and Enrico Clementi
IBM Corporation, P.O. Box 390,
Poughkeepsi, New York 12602

I. INTRODUCTION

Despite the availability of fast computers for the last ten years, computations of Hartree-Fock quality wavefunctions (wf) for systems with more than 30 to 40 atoms are nonexistent, and very few examples of computations with 50 or more atoms are available at the minimal basis set level.

In this work we analyze the statistical distribution of the matrix elements one encounters in chemical systems with many atoms. Such distributions are intuitively expected to be different from those of a system with the same number of elements, but much fewer, heavier atoms. The analyses is performed with reference to few examples where a small and highly contracted gaussian basis set is used. This contracted basis set is similar to a Slater minimal basis set[1]; thus our conclusion should hold also for the latter. The aim of this work is to examine the possibility of introducing algorithms useful to reduce the computer time of the computation without loss of accuracy. Most of the analyses below reported is based on data obtained from an SCF computation of a phosphate, two sugar units and a base (guanine), S-P-S-G for short; the geometries are standard geometries for the B-DNA double helix[2]. The fragment considered consists of 25 second row atoms (C,N,O), a phosphorous atom and 16 hydrogen atoms. The basis set is a (7s/3p) primitive set for C, N and O contracted to minimal with 5 s functions in 1s, 2 s functions in the 2s. The phosphorous set is (9s/6p) contracted to minimal with 5, 2 and 2 primitives in the 1s, 2s and 3s respectively and 3 primitives in each of the 2p and 3p. For the hydrogen atoms four primitives are contracted in one function (for more details, see elsewhere)[3]. These basis sets yield atomic energies somewhat superior to those obtained with minimal basis sets of Slater functions. The structure of the S-P-S-G molecule is reported below.

C. A. Weatherford and H. W. Jones (eds.), ETO Multicenter Molecular Integrals, 157–169.

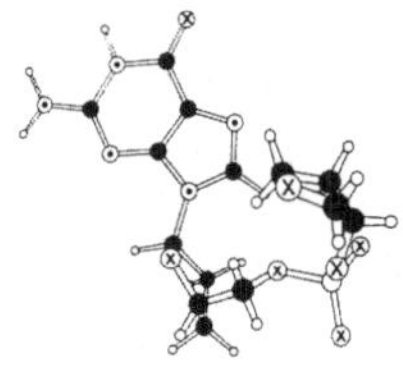

In a chemical system with many interacting atoms the electronic distribution extends over a considerable volume, notably larger than the volume within which a single basis function spreads. Therefore, many functions will have no interactions with each other, and, therefore, one expects a large number of very small matrix elements[4]. Because in the limit of a large number of interacting atoms there is a statistically uniform distribution of atoms in space, a quasi continuous statistical distribution of large to small matrix elements can be expected. In this work we want to examine this point in some detail; the conclusions we shall reach might be used as the starting point for a molecular program specifically oriented to large molecules. The latter task is not discussed in this work.

II. DISTRIBUTIONS FOR THE MATRIX ELEMENTS IN P-S-G MOLECULE

As our basis functions are centered on the atoms all two electron matrix elements can be divided in different groups according to whether the functions are on the same center or not. Seven different types are possible (AA|AA), (AA|AB), (AB|AB), (AA|BB), (AA|BC), (AB|AC) and (AB|CD) where the notation A, B, C, D is a short notation for both the function and its center. This structure is the one used for example since IBMOL5 and subsequent versions[5]. The analyses here adopted is obtained for integrals computed with IBMOL7. The following analyses is restricted to group functions, namely frozen linear combination of primitive gaussians (i.e., contracted gaussian functions).

In the figures we report the number of integrals (below specified) with a value included in the interval $10^{\log}-0.2$ and $10^{\log}-0.3$. The figures show the number $\mathcal{N}$ of integrals of a given value (ordinate), specified as decimal algorithm on the abscissa. Inset A reports the overlap integrals $|S_{ij}|$ as absolute values, inset B reports the closed shell Hartree-Fock density matrix elements (the a,b and c diagrams correspond to the absolute value, the positive value and the negative value distributions respectively); inset C reports in diagram a the products $|S_{ij}\ S_{kl}|$ corresponding to four center two electron integrals (AB|CD). Inset D reports the distribution for the analysis for three additional large systems in order to give a somewhat more general basis of our results; the systems are explicitly indicated at the top of Figure 5. System A is obtained from system S-P-S-G by subtracting the guanine unit, system B is benzopyrene an almost planar molecule and system C is a small molecule specifically a Zn++ ion coordinated to two ammonia molecules and one Oh- radical but treated in a larger basis.

III. ANALYSES OF THE DISTRIBUTION

In discussing distributions it is convenient to introduce an adjoined function δ_{iA} for each group function χ_{iA}(centered on A); the adjoined function can be approximatively assumed such that $\langle\delta_{iA}|\chi_{iA}\rangle$ is the nearest number to unity. We call α_{iA} the exponent of the adjoined function.

The overlap integrals distribution is composed of three parts as obvious by considering $\langle\chi_{iA}|\chi_{jB}\rangle$. There are N diagonal elements with unit value, $1/2 \sum_{A=1}^{M} N(A)\,(N(A)-1)$ one center non-diagonal elements $\langle\chi_{iA}|\chi_{jA}\rangle$ with $i \neq j$ and $N(N-1)-1/2 \sum_{A=1}^{M} N(A)^2$ two center non-diagonal elements. The one center non-diagonal elements are proportional to $[\pi/(\alpha_{iA}+\alpha_{jA})]^{3/2}$ and the two center non-diagonal elements are proportional to $[\pi/(\alpha_{iA}+\alpha_{jB})]^{3/2} \exp\left(-R^2_{AB}\,\frac{(\alpha_{iA}\cdot\alpha_{jB})}{\alpha_{iA}+\alpha_{jB}}\right)$ where R_{AB} is the distance between the centers A and B. In general 2 a.u. $<R^2<\infty$. In the inset a of Figure 1, we can note the peak of the N diagonal elements containing also non-diagonal elements on its right side followed by the remaining one center non-diagonal elements, partly over imposed (initial small shoulder) to the distribution of the two centers non-diagonal elements. The first two peaks on the left side of the distribution correspond to the functions located on nearest neighbors. The latter distribution has an exponential decay and is bound by the maximum and minimum value assumed by $(\alpha_{iA}\cdot\alpha_{jB})/(\alpha_{iA}+\alpha_{jB})$. For a very large molecule the distribution of the R_{AB} is expected to be more regular. In our case, the modest number of values (528) precludes a more accurate analyses of general validity for a large molecule (with about 100 to 200 atoms). The integrals analyzed in inset A are = 11325. The presence or absence of a smooth distribution of overlap matrix elements can be taken as an indirect definition of "large" or "small" molecules. Alternatively, and more correctly, a nearly regular distribution of R_{AB} distances (the inter-nuclear distance) can be taken as a necessary condition for a "large" molecule.

In insets C and D, we report distributions of products of overlap integrals, being those relevant to the Mulliken's integral approximation for four centers. In (a) of C and in (a) and (b) of D we report the distributions for $|S_{AB}\,S_{CD}|$, $|S_{AA}\,S_{BC}|$ and $|S_{AB}\,S_{AC}|$, respectively. Notice that the value of $\mathcal{N}$ goes to many thousands of numbers, thus, these distributions have a good statistical value. Notice in addition the peak by peak similarity of the $|S_{AA}\,S_{BC}|$ distribution with the $|S_{ij}|$ distribution; this is due to the fact that S_{AA} distribution is almost the δ distribution, namely 150 $\delta(S_{iAjA}-1)$. Notice in addition

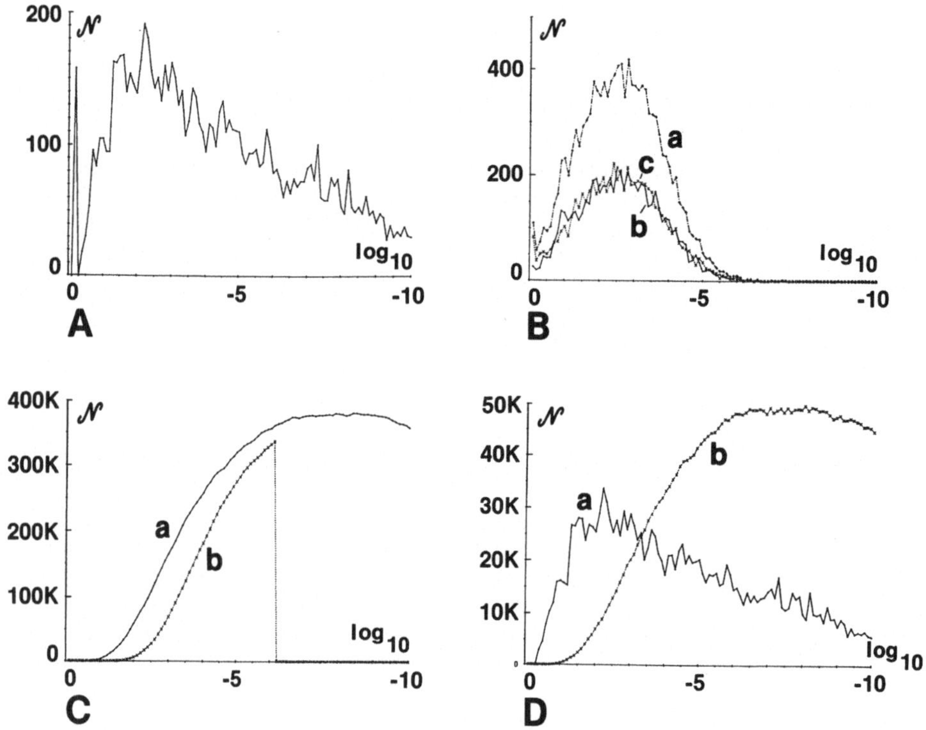

Fig. 1. Statistical distribution of integrals.

The number $\mathcal{N}$ of integrals whose value lies between $10^{\log-0.2}$ and $10^{\log-0.3}$ is plotted versus log.

A. Overlap integrals $|S_{ij}|$

B. Density matrix elements D_{ij} , a) $|D_{ij}|$, b) $D_{ij}>0$, c) $D_{ij}<0$

C. a) $|S_{AB}S_{CD}|$ b) four center integrals $|(AB|CD)|$

D. a) $|S_{AA}S_{BC}|$ b) $|S_{AB}S_{AC}|$

the similarity of the $|S_{AB}\ S_{CD}|$ and $|S_{AB}\ S_{AC}|$ distributions (curve a in C and b in D). In inset C the distribution of the four center integrals $|(AB|CD)|$ is reported (diagram a) with a cutoff at 10^{-6} a.u. The overlap product distribution and the integral distributions are notably similar. They almost match, if the integrals are multiplied by 7, an average distance between the atoms in S-P-S-G. This suggests that there should be the possibility to reformulate the Mulliken approximation.

In inset B the density matrices are reported; the interesting features are the nearly gaussian distribution and the nearly equal partitioning into populations with positive and negative sign. It is noted that this situation is not expected to be true for open shell cases and excited states.

These regularities observed for the density matrix elements and for the products of overlap integrals are expected to bring about regularities in the two-electron integrals in particular in the three and especially in the four center cases. The two electron integrals of one center and two center type (Figure 2, inset A, B and C) are not considered in detail, since it is reasonable to compute such integrals accurately. We note however that the first peak in inset B (curve a) corresponds to the very long range positive coulomb integrals $(ii|ll)$ -about 20000- which are not so many as to require approximations. The nice cancellation (due to opposite signs) for smaller integrals reflects the equal distribution of signs in the hybrid densities. One can notice, that the distributions for each one of the three types $(AA|AB)$, $(AB|AB)$ and $(AA|BB)$ (insets B and C) are characteristic and physically meaningful. In all these diagrams the distribution is truncated for elements smaller than 10^{-6} a.u.

Concerning the four center integrals we wish to point out the importance of the threshold selection for the integral computation. The total number of four center two electron integrals is 25×10^6, 7.4×10^6, 4.3×10^6 and 1.8×10^6 with thresholds 0 a.u., 10^{-6} a.u., 10^{-5} a.u. and 10^{-4}a.u., respectively. Namely an increase of one order of magnitude in the threshold reduces the number of four center integrals by a factor of 2. If one neglects all integrals smaller than 10^{-6}, there is no need to calculate integrals of the absolute value of 10^{-5} with an accuracy of 8 digits; one digit is sufficient[5,6]. It is noted that approximations reliable to only one or two digits could reduce appreciably the computing time.

We now pass to the next aspect of our analyses, namely to the product of two electron integrals times the density matrix elements. The two electron integrals and density matrix elements yield coulomb and exchange energy contributions which are here analyzed in terms of distributions. In Figure 3 the one and two center coulomb energies are reported in insets A and B respectively; inset C considers the three centers and inset D the four centers. Three different cut offs for the two electron integrals have been selected at 10^{-4}, 10^{-5} and 10^{-6} a.u. The positive and the negative values of the products $(AB|CD)\ D_{AB}\cdot D_{CD}$

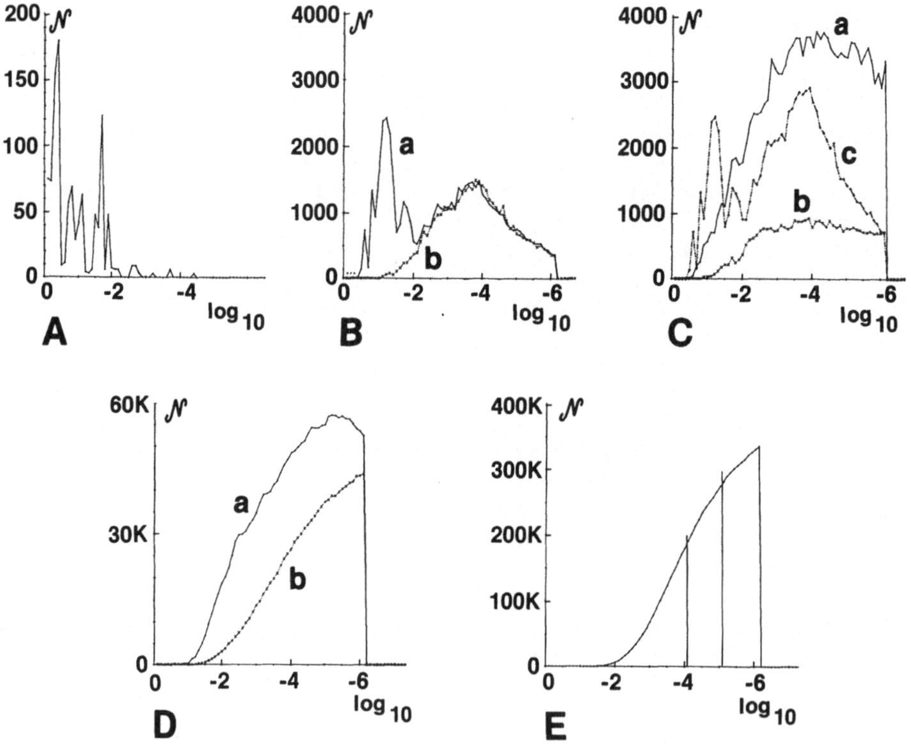

Fig. 2. Two electron integrals $(ij \mid kl)$

A. type $|(AA \mid AA)|$

B. a) $(AA \mid BB) > 0$ b) $(AA \mid BB) < 0$

C. a) $|(AA \mid AB)|$ b) $|(AB \mid AB)|$

c) $|(AA \mid BB)|$

D. a) $|(AA \mid BC)|$ b) $|(AB \mid AC)|$

E. $|(AB \mid CD)|$

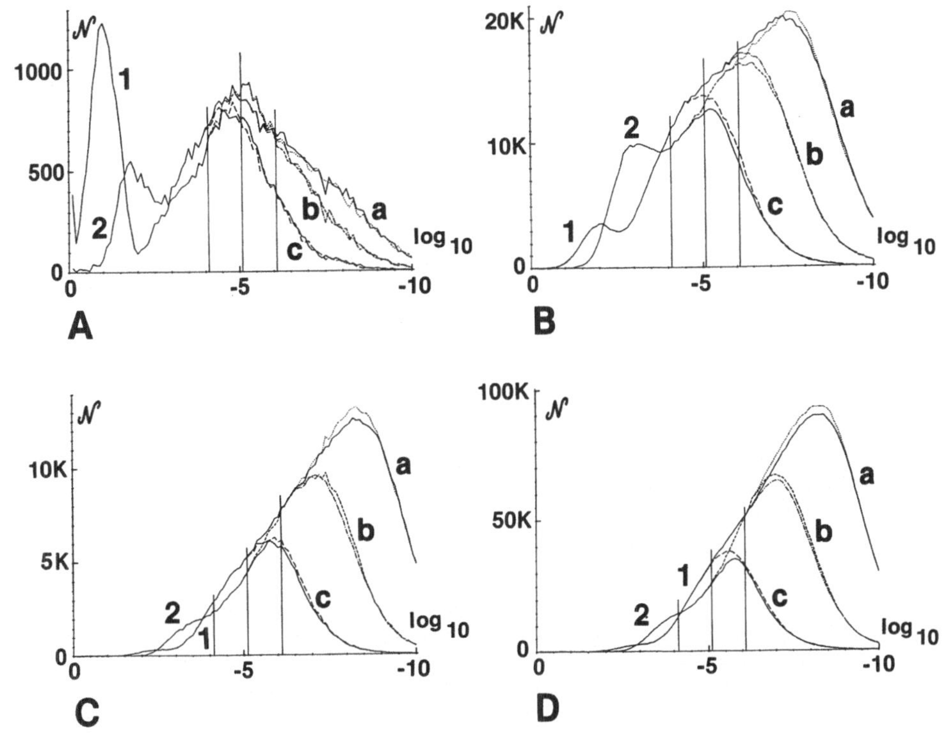

Fig. 3. Coulomb energy contribution $j_{ijkl}=D_{ij}D_{kl}\ (ij|kl)$

A. $j_{AAAA} + j_{AABB}$ B. $j_{AAAB} + j_{AABC}$

C. $j_{ABAB} + j_{ABAC}$ D. j_{ABCD}

1) $j > 0$ 2) $j < 0$

with integrals cutoff

a) 10^{-6} b) 10^{-5} c) 10^{-4} , respectively.

are differentiated with the index 1 (positive) and 2 (negative). The first obvious observation is that for small values the positive and the negative distrbutions nearly cancel out exactly. The second observation is that in the large value region, the negative integral population exceeds the positive integral population; at smaller values this situation is reversed. The curves describing the distributions for different thresholds coincide for those values which are larger than the integral cutoff, since the density matrix elements are seldom much larger than one. Thus from inset A it can be seen that they are becoming different at their integral cutoff.

But when the integrals are multiplied by two center density matrix elements (one "two center" density matrix in inset B and two in C and D) the curves coincide for a half or a whole order of magnitude beyond their integral cutoff. Therefore, the energy contributions are half an order of magnitude more accurate than the integrals. This suggests to select different thresholds for different types of integrals. Finally, because of the very large number of small integrals one might assume an "a priory" distribution and explicitly compute only the large value integrals and analytically extrapolate the remaining ones. Work is in progress along these directions.

By considering the four center exchange energies (see Figure 4 inset B, diagram C) the cancellation of the positive and negative integrals becomes nearly exact and remains notably accurate even for the three center exchange (see Figure 4 inset B, diagrams a and b). These rather spectacular regularities carry over in the total energy contributions reported in inset C of Figure 4. As previously often noted in energy partitioning[7], the main contribution is associated with two centers, followed by three and four center energies. This conclusion becomes obvious by inspection of inset C of Figure 4. However, here we learn of the existence of an unexpected regular pattern. Despite the fact that the number of points constituting the curves is relatively small the distribution seems to be a continuous function, with an analytical description.

IV. ADDITIONAL EXAMPLES

Before concluding, we should ask how general are the reported regularities in the distributions. For this reason we report in Figure 5 some of the data obtained with three different molecules. The first is designated as S-P-S and it is a smaller fragment of S-P-S-G where the guanine molecule has been replaced with a terminal hydrogen atom. The second example benzopyrene is a very different molecule because mainly planar. The third example is for a small molecule with however many electrons due to the presence of a Zn atom. For the S-P-S and for benzopyrene we have used the basis set described for S-P-S-G. For the Zn complex we have selected an extended basis set of double zeta polarization quality. We report the density distribution, the four center coulomb and exchange energies. From these insets it seems possible to

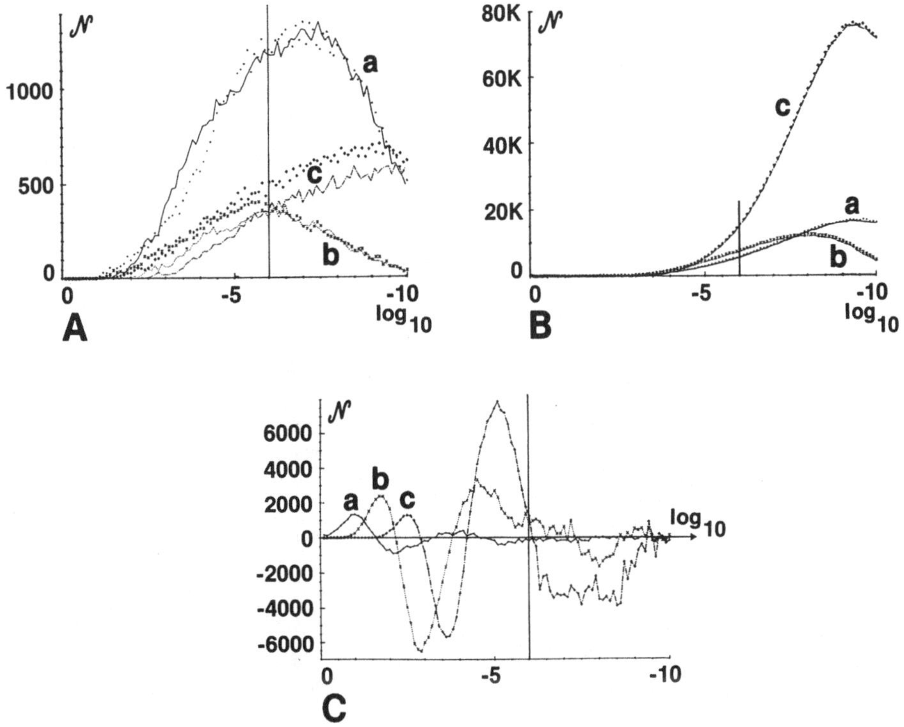

Fig. 4.

A. and B. exchange energy contributions

$$k_{ijkl} = 1/4\ (D_{il}\ D_{jk} + D_{ik}\ D_{jl})\ (ij \mid kl)$$

Dotted lines refer to $k_{ijkl} > 0$, full lines to $k_{ijkl} < 0$

A. a) k_{AAAB} b) k_{ABAB} c) k_{AABB}

B. a) k_{AABC} b) k_{ABAC} c) k_{ABCD}

C. Two electron closed shell Hartree-Fock energy contributions

$$E_{ijkl} = j_{ijkl} - k_{ijkl}$$

a) $E_{AAAB} + E_{ABAB} + E_{AABB}$ b) $E_{AABC} + E_{ABAC}$ c) E_{ABCD}

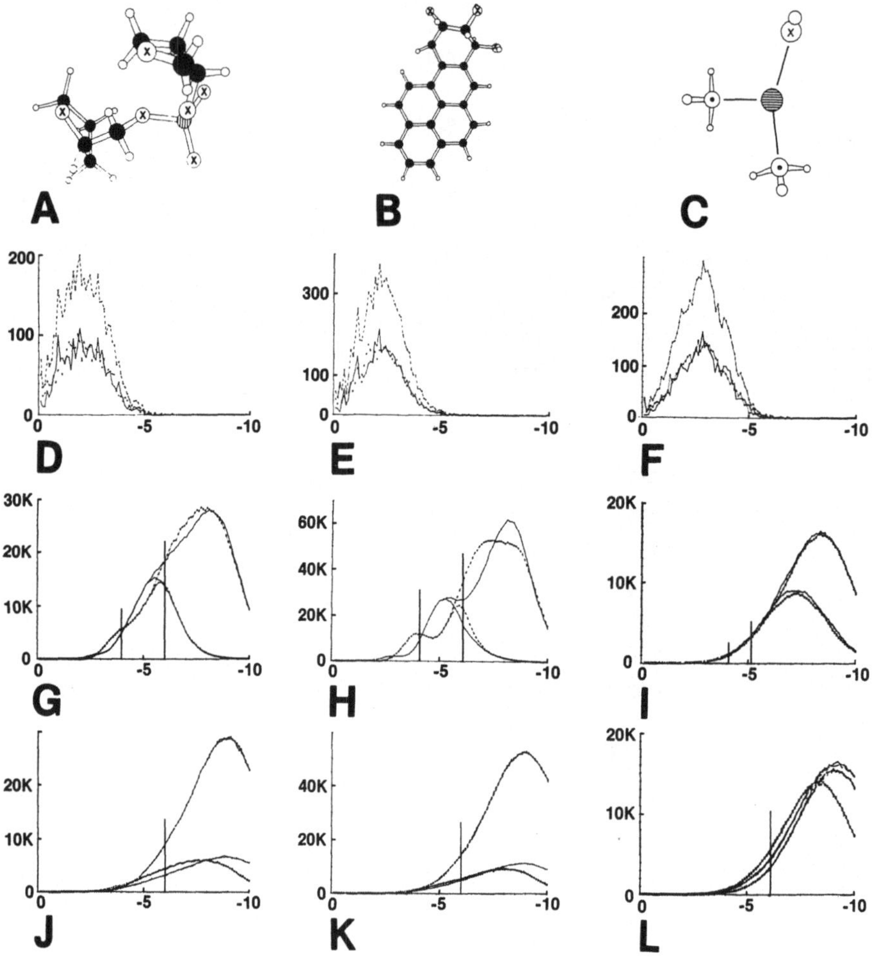

Fig. 5. Comparison of three different systems

A-C. The structure of the systems

• carbon, o hydrogen, ⓧ oxygen, ⊙ nitrogen, zinc, phosphorus

A) sugar - phosphate - sugar (SPS)

B) Benzopyrene (BP), C) $(Zn\,(NH_3)_2OH))^+$

D-F. D_{ij} details in Fig. 1 inset B

G-I. j_{ijkl} details in Fig. 3 inset D

J-L. k_{ijkl} details in Fig. 4 inset B

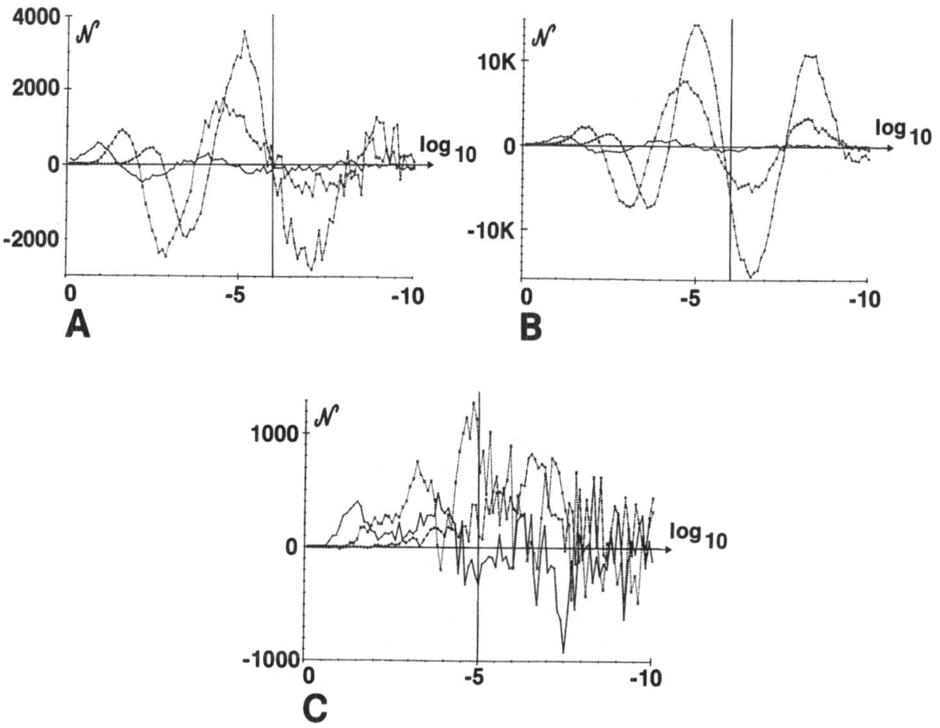

Fig. 6. Energy Contributions E : for details see Fig. 4 inset C

A. SPS, B. BP, C. Zn

assume that the distributions presented for S-P-S-G are general distributions and the conclusions seem to hold also for these cases.

In Figure 6, we consider the total energies for S-P-G, for benzopyrene and for the Zn complex. Again we conclude that the distributions discussed for S-P-S-G are likely generally valid for large molecules, namely molecules with many atoms. However our conclusions do not have to be valid for small molecules with many electrons (see the Zn complex) or for extended basis sets.

V. CONCLUSIONS

In general today large molecules are computed with algorithms that are a direct extension of those introduced for smaller molecules in the middle 1960's. The improvements added since that time have been rather "marginal," even if important in gaining computational speed. With chemical systems composed of many atoms we have shown that there are regular distributions. These regularities should help in devising numerical techniques specifically adapted for large molecules. The essential conditions for the regularity above reported must be implicit in the regularities of the density matrices distributions and in the sparse matrix character of the integrals. In turn these regularities are the effect of the nearly continuous distribution of the overlap matrix elements and therefore of the R_{AB} distributions in a large molecule. These regularities are very clear especially in the three and four center matrix elements, the most time consuming section in a large molecule computation. The very tentative suggestion one can advance at this preliminary stage of this work is that matrix elements involving nearest neighbors should be analitically computed (as standardly done). Matrix elements for non nearest neighbors might be approximated not on a one by one basis but in a cumulative way, taking eventually advantage of the distribution, if these can be analytically represented.

ACKNOWLEDGMENT

The integrals and wave functions here analyzed have been obtained from Dr. G. Corongiu. We acknowledge the use of such data.

REFERENCES

1. Clementi, E. and Raimondi,D.L.: 1963, J. Chem. Phys. 38, p.2686; Clementi, E., Raimondi,D.L. and Reinhardt, W.R.: 1967, J. Chem. Phys. 47, p. 1300.

2. Fieldman, R.: 1976, Atlas of Macromolecules, Document 13.2.1.1.1, Natl. Inst. Health, Bethesda, Maryland, U.S.A.

3. Gianolio, L., Pavani, R. and Clementi, E.: 1978, Gass. Chim. Ital. 108, p.181.

4. Clementi, E.: 1972, Proc. Nat. Acad. Sci. U.S.A. 69, p.2942.

5. Clementi, E. and Mehl, J.: June 22, 1971, IBMOL5, Program Manual, IBM Tech. Report; see also Ortoleva, Castellani and Clementi, E.: 1980, J. Comp. Phys. 19, p.337.

6. Clementi, E.: 1972, Selected Topics in Molecular Physics, Verlag Chemie, p.199; Clementi, E. and Mehl, J.: 1974, Jerusalem Symposia in Quantum Chemistry and Biochemistry VI, Academic Press, p.137.

7. Clementi, E.: 1967, J. Chem. Phys. 46, p.3842; Clementi, E.: 1976, Lecture Notes in Chemistry Vol. 2, 1980, Lecture Notes in Chemistry Vol. 19, Springer-Verlag.

USE OF THE NEUMANN EXPANSION IN EVALUATION OF MULTICENTER ELECTRON-REPULSION INTEGRALS FOR SLATER-TYPE ORBITALS

I.I. Guseinov
Physical Faculty, Azerbaijan State University, 23 Lumumba Street, Baku, USSR

Convenient formulas are derived for molecular two-center Coulomb, hybrid and three- and four-center electron-repulsion integrals by use of the Neumann expansion for the inverse distance $\frac{1}{r_{12}}$ in the elliptical coordinates which occur in linear-combination-of-atomic orbitals calculations with Slater-type orbitals (STO's). The final results are expressed in terms of the well-known auxiliary functions $q_k(P,\mu)$. The convergence of the series is tested by calculating concrete cases.

In recent years it has become recognized that an efficient and accurate method of evaluating the multicenter integrals is very essential for extracting meaningful conclusion on the electronic structure of molecules from detailed calculations using exact values of all integrals between STO's. For this purpose several methods have been reported in the literature (the spherical harmonic expansion methods for STO's about an arbitrary point[1-3], Gaussian transform methods[4], Fourier transform methods[5-10] and the method using the reduced Bessel functions[11-12]). In previous publications[13-15] we have presented a unified treatment of all the multicenter electron-repulsion integrals without the use of the Neumann expansion, except the two-center exchange integrals: with the help of single- and two-center expansion methods without the use of the Neumann expansion these integrals have been expressed in terms of the usual A and B functions[15] and the auxiliary functions $\bar{G}^{g}_{-NN'}$[14], respectively. For the calculation of the functions $\bar{G}^{g}_{-NN'}$ useful recurrence relationships have been established in our works[13,16].

In this paper, by use of the Neumann expansion, the general formulas are obtained for multicenter electron-repulsion integrals in terms of the auxiliary functions $q_k(P,\mu)$ which can be calculated from the recurrence relationships[17]. In (17) with the help of Neumann expansion we have derived completely general expressions for two-center exchange integrals and three-center nuclear attraction integrals. For the calculation of two-center coulomb and hybrid integrals we use the method set out in this work. Then we obtain for these integrals the

C. A. Weatherford and H. W. Jones (eds.), ETO Multicenter Molecular Integrals, 171–176.

following infinite series over ℓ:

Coulomb integrals

$$\left[(n_a \ell_a \sigma_a)(n'_a \ell'_a \sigma'_a) \,\middle|\, (n_\ell \ell_\ell \sigma_\ell)(n'_\ell \ell'_\ell \sigma'_\ell)\right] =$$

$$\frac{2}{R} N_{n_a n'_a}(P_a,t_a) N_{n_\ell n'_\ell}(P_\ell,t_\ell) \sum_{\ell=0}^{\infty} \sum_{m=-\ell}^{\ell} A^m_{\sigma_a \sigma'_a} A^m_{\sigma_\ell \sigma'_\ell} (-1)^\lambda (2\ell+1) \times$$

$$\left[\frac{(\ell+\lambda)!}{(\ell-\lambda)!}\right]^2 \sum_{L_a L_\ell} \left[(2L_a+1)(2L_\ell+1)\right]^{\frac{1}{2}} C^{L_a\lambda}(\ell_a\lambda_a,\ell'_a\lambda'_a) \times$$

$$C^{L_\ell\lambda}(\ell_\ell\lambda_\ell,\ell'_\ell\lambda'_\ell) \sum_{\alpha_a q_a s_\alpha \beta_\ell q_\ell s_\ell} g^{q_a}_{\alpha_\alpha o}(L_a\lambda,00;0) F_{s_a}(n_a+n'_a-1-\alpha_a,1) \times$$

$$B^{\ell\lambda}_{s_a+q_a}(P_a)\, g^{q_\ell}_{o\beta_\ell}(00,L_\ell\lambda;0) F_{s_\ell}(1,n_\ell+n'_\ell-1-\beta_\ell)\, B^{\ell\lambda}_{s_\ell+q_\ell}(-P_\ell) \times$$

$$\Omega^{\ell\lambda,\ell\lambda}_{n_a+n'_a-\alpha_a-s_a+q_a,\; n_\ell+n'_\ell-\beta_\ell-s_\ell+q_\ell;\; n_\ell+n'_\ell-\beta_\ell-s_\ell+q_\ell,\; n_a+n'_a-\alpha_a-s_a+q_a,}(P_a,P_\ell) \tag{1}$$

Hybrid integrals

$$\left[(n_a \ell_a \sigma_a)(n'_a \ell'_a \sigma'_a) \,\middle|\, (n''_a \ell''_a \sigma''_a)(n_\ell \ell_\ell \sigma_\ell)\right] =$$

$$\frac{2}{R} N_{n_a n'_a}(P_a,t_a) N_{n''_a n_\ell}(P,t) \sum_{\ell=0}^{\infty} \sum_{m=-\ell}^{\ell} A^m_{\sigma_a \sigma'_a} A^m_{\sigma''_a \sigma_\ell} (-1)^\lambda (2\ell+1) \times$$

$$\left[\frac{(\ell+\lambda)!}{(\ell-\lambda)!}\right]^2 \sum_{L_a} (2L_a+1)^{\frac{1}{2}} C^{L_a\lambda}(\ell_a\lambda_a,\ell'_a\lambda'_a) \times$$

$$\sum_{\alpha_a q_a s_a, \alpha\beta q s} g^{q_a}_{\alpha_\alpha o}(L_a\lambda,00;0) F_{s_a}(n_a+n'_a-1-\alpha_a,1)\, B^{\ell\lambda}_{s_a+q_a}(P_a) \times$$

$$g^q_{\alpha\beta}(\ell''_a\lambda''_a,\ell_\ell\lambda_\ell;\Lambda) F_s(n''_a-\alpha,n_\ell-\beta)\, B^{\ell\lambda}_{s+q}(Pt) \times$$

Table 1. Two-center carbon monoxide hybrid and exchange integrals and their convergence as a function of number of summation terms.

Number of terms / Integrals	0	1	2	3	4	5	6	Comparison Values (18)
$2s_C\ 2s_C \| 2s_C 1s_O$	0,031458	0,023151	0,022633	0,022905	0,022890	0,022892		0,02289
$2p\Pi_C\ 2p\Pi_C \| 1s_C 2s_O$	0,029003	0,036633	0,036614	0,036515	0,036530	0,036534		0,03653
$2p\sigma_C\ 2p\sigma_C \| 2p\sigma_C\ 2p\sigma_O$	0,207792	0,211908	0,212031	0,213759	0,214105	0,214112	0,214116	0,21412
$2p\Pi_C\ 2p\sigma_C \| 2p\Pi_C\ 2p\sigma_O$	0,106082	0,010687	0,008782	0,010663	0,010685	0,010699		0,01070
$2p\Pi_C\ 2p\Pi_C \| 2p\Pi_C\ 2p\Pi_O$	0,163815	0,174370	0,171144	0,171855	0,171758	0,171734	0,171732	0,17173
$2p\Pi_C\ 2p\bar{\Pi}_C \| 2p\Pi_C\ 2p\bar{\Pi}_O$	-	-	0,003161	0,007519	0,007360	0,007299	0,007296	0,00730
$2s_C\ 2s_O \| 2s_C\ 1s_O$	0,016922	0,018850	0,017708	0,017599	0,017603	0,017601		0,01760
$2p\Pi_C\ 2p\Pi_O \| 1s_C\ 2s_O$	0,008957	0,008308	0,007936	0,007951	0,007949			0,00795
$2p\sigma_C\ 2p\sigma_O \| 2p\sigma_C\ 2p\sigma_O$	0,106082	0,107084	0,130308	0,131419	0,131569	0,131588	0,131589	0,13159
$2p\Pi_C\ 2p\sigma_O \| 2p\Pi_C\ 2p\sigma_O$	0,106082	0,014064	0,015565	0,016038	0,016040	0,016041		0,01604
$2p\Pi_C\ 2p\Pi_O \| 2p\Pi_C\ 2p\Pi_O$	0,040643	0,041195	0,044769	0,044937	0,044942			0,04494
$2p\Pi_C\ 2p\bar{\Pi}_O \| 2p\Pi_C\ 2p\bar{\Pi}_O$	-	-	0,002265	0,002353	0,002355	0,002356		0,00236

$$\Omega^{\ell\lambda,\ell\lambda}_{n_a+n'_a-\alpha_a-s_a+q_a,\, n''_a-\alpha+n_\ell-\beta-s+q;\, n''_a-\alpha+n_\ell-\beta-s+q,\, n_a+n'_a-\alpha_a-s_a+q_a}(P_a,P) \qquad (2)$$

The calculation of these integrals and two-center exchange integrals on a computer shows that convergence of series is rapid, therefore it is sufficient to take into account a few number of terms in the formulas of two-center electron-repulsion integrals. We can test the convergence of the series by calculating concrete cases. The results of tests on series accuracy and the comparison values[18] are given in Table 1 for several two-center carbon monoxide hybrid and exchange integrals. As we can see from this table, the accuracy was assessed by increasing the number of terms taken in the infinite summations. An overall accuracy 10^{-6} hartree was sought for 6 terms. Greater accuracy is easily attainable by the use of more terms of the expansion.

It should be noted that, if the effective nuclear charges of the atomic orbitals $(n''_a\ell''_a\sigma''_a)$ and $(n_\ell\ell_\ell\sigma_\ell)$ in (2) are the same (t=0), the infinite series in the formula of two-center hybrid integrals breaks off after certain terms. In this case upper limit of the sum depends on the quantum numbers of the atomic orbitals $(0\leq\ell\leq n''_a+n_\ell-\lambda)$.

Now we can move on to the calculation of three-and four-center electron-repulsion integrals with the aid of the Neumann expansion. For this purpose we use the expansion of STO's about a point displaced from the orbital center[15]. Then we obtain for these integrals the following formulas in terms of the overlap integrals, the two-center hybrid or exchange integrals:

three-center hybrid integrals

$$\left[(n_a\ell_a m_a)(n'_a\ell'_a m'_a)\,\middle|\,(n_c\ell_c m_c)(n_\ell\ell_\ell m_\ell)\right] =$$

$$\sum_{n''_a\ell''_a m''_a,\, n'''_a} M^{n_c-1,\ell''_a}_{n_c\ell_c m_c,\, n''_a\ell''_a m''_a}(P_{ca''},t_{ca''})\, \omega^{\ell''_a}_{n''_a n'''_a} \times$$

$$\left[(n_a\ell_a m_a)(n'_a\ell'_a m'_a)\,\middle|\,(n'''_a\ell''_a m''_a)(n_\ell\ell_\ell m_\ell)\right] \quad \text{for } \zeta''_a=\zeta_\ell \qquad (3)$$

three-center exchange integrals

$$\left[(n_a\ell_a m_a)(n_\ell\ell_\ell m_\ell)\,\middle|\,(n_c\ell_c m_c)(n'_\ell\ell'_\ell m'_\ell)\right] =$$

$$\sum_{n'_a \ell'_a m'_a,\, n''_a} M^{n_c-1,\ell'_a}_{n_c \ell_c m_c,\, n'_a \ell'_a m'_a}(P_{ca'}, t_{ca'})\, \omega^{\ell'_a}_{n'_a n''_a}$$

$$\Big[(n_a \ell_a m_a)\ (n_\ell \ell_\ell m_\ell)\,\Big|\,(n''_a \ell'_a m'_a)\ (n'_\ell \ell'_\ell m'_\ell)\Big] \qquad \text{for } \zeta'_a = \zeta'_\ell \qquad (4)$$

four-center integrals

$$\Big[(n_a \ell_a m_a)(n_c \ell_c m_c)\,\Big|\,(n_d \ell_d m_d)\ (n''_\ell \ell''_\ell m''_\ell)\Big] =$$

$$\sum_{n_\ell \ell_\ell m_\ell, n'_\ell} M^{n_c-1,\ell_\ell}_{n_c \ell_c m_c,\, n_\ell \ell_\ell m_\ell}(P_{c\ell}, t_{c\ell})\, \omega^{\ell_\ell}_{n_\ell n'_\ell} \sum_{n''_a \ell''_a m''_a,\, n'''_a} \times$$

$$M^{n_d-1,\ell''_a}_{n_d \ell_d m_d,\, n''_a \ell''_a m''_a}(P_{da''}, t_{da''})\ \omega^{\ell''_a}_{n''_a n'''_a} \times$$

$$\Big[(n_a \ell_a m_a)(n_\ell \ell_\ell m_\ell)\,\Big|\,(n'''_a \ell''_a m''_a)\ (n''_\ell \ell''_\ell m''_\ell)\Big] \qquad \text{for } \zeta_\ell = \zeta_a,\ \zeta''_a = \zeta''_\ell \qquad (5)$$

where $P_{ca} = \dfrac{R_{ca}}{2}(\zeta_c + \zeta_a)$, $t_{ca} = \dfrac{\zeta_c - \zeta_a}{\zeta_c + \zeta_a}$.

It should be noted that the two-center hybrid and exchange integrals which occur in (3), (4) and (5) are finite sums and determined from (2) and the formulas of Ref.(17), respectively. Therefore, we have the infinite series only from the expansion of STO's about a point displaced from the orbital center the convergence of which has been tested in (15).

REFERENCES

1. Barnett, M.P. and Coulson, C.A.: 1951, Phil. Trans. R. Soc. London Ser. A 243, p.221.

2. Harris, F.E. and Michels, H.H.: 1965, J. Chem. Phys. 43, p.165; 1966, 45, p.116.

3. Sharma, R.R.: 1976, Phys. Rev. A 13, p.517.

4. Shavitt, I. and Karplus, M.: 1965, J. Chem. Phys. 43, p.398.

5. Bolotin, A.B. and Shugurov, V.K.: 1963, Zh. Vychisl. Mat. Mat. Fiz. 3, p.560.

6. Rakauskas, R. and Bolotin, A.: 1965, Leit. Fiz. Rinkinys 5, p.305.

7. Bolotin, A.B. and Rakauskas, R.: 1965, Liet. Fiz. Rinkinys 5, p.473.

8. Rakauskas, R.,Poshunaite, H.P. and Bolotin, A.B.: 1968, Liet. Fiz. Rinkinys 8, p.107.

9. Silverstone, H.J.: 1968, J. Chem. Phys. 48, pp.4098,4106.

10. Kay, K.G. and Silverstone, H.J.: 1969, J. Chem. Phys. 51, p.4287.

11. Filter, E. and Steinborn, E.O.: 1978, Phys. Rev. A 18, p.1.

12. Steinborn, E.O. and Filter, E.: 1980, Int. J. Quantum Chem. 18, p.219.

13. Guseinov, I.I.: 1977, J. Chem. Phys. 67, p.3837.

14. Guseinov, I.I.: 1978, J. Chem. Phys. 69, p.4990.

15. Guseinov, I.I.: 1980, Phys. Rev. A 22, p.369.

16. Guseinov, I.I.: (in press), J. Chem. Phys.

17. Guseinov, I.I.: 1976, J. Chem. Phys. 65, pp.4718,4722.

18. Hurley, A.C.: 1960, Revs. Modern Phys. 32, p.400.

DISPERSION ENERGY FROM ATOMIC FORM FACTORS

Ker-Fong Lee
Department of Chemistry, Florida State University, Tallahassee, Florida 32306, USA

A simple numerical scheme for the evaluation of the integrals encountered in the calculations of the second order dispersion energy for H-H and He-He diatoms using static form factors.

1. INTRODUCTION

The charge density susceptibility formulation of inter-molecular forces provides a useful framework for treating molecules of arbitrary sizes and electron delocalizations whose electron exchange effects are neglected[1]. The susceptibility is a nonlocal polarizability which takes account of spatial as well as temporal correlation. In a recent publication[2], Linder, Lee, Malinowski and Tanner applied the Unsöld approximation[3] to the susceptibilities of totally uncorrelated wavefunctions as well as a single determinant two electron function expressed entirely in terms of the experimentally measurable and theoretically calculable static form factors F(k). The theory was applied to the second order dispersion interaction of two helium atoms using the very accurate analytic wavefunction reported by Clementi and Roetti[4], and later extended to include the second order polarization interaction[5]. In both applications, all the integrals were evaluated analytically[6].

We are reporting here a numerical scheme for evaluating those integrals. Our computer codes simply read as input data the form factors of the interacting atoms and then evaluate the integrals numerically. We perform calculations on the interactions of the H-H and He-He diatoms as a function of their internuclear distances R. For the sake of conciseness, we shall confine our treatment to the dispersion energy only since inclusion of the polarization energy may be similarly accomplished.

2. PROCEDURE

The second order dispersion energy in Unsöld approximation is[2],

C. A. Weatherford and H. W. Jones (eds.), ETO Multicenter Molecular Integrals, 177–180.

$$E^{(2)}_{disp}(R) = \frac{-e^4}{2\bar{\varepsilon}} \left\{ \frac{2IFL}{N} + IFF + \frac{ILL}{N^2} \right\} \tag{1}$$

where N is equal to 1 for hydrogen and 2 for helium, $2\bar{\varepsilon}$ the mean excitation energy of the diatom, e the electronic charge and

$$IFL = -In\ F(|\underset{\sim}{k}-\underset{\sim}{k}'|)F(-\underset{\sim}{k})F(\underset{\sim}{k}') \tag{2}$$

$$IFF = In\ F(|\underset{\sim}{k}-\underset{\sim}{k}'|)F(|\underset{\sim}{k}'-\underset{\sim}{k}|) \tag{3}$$

$$ILL = In\ F(\underset{\sim}{k})F(-\underset{\sim}{k}')F(-\underset{\sim}{k})F(\underset{\sim}{k}') \tag{4}$$

in which

$$In = \frac{1}{4\pi^4} \int_{-\infty}^{\infty} d\underset{\sim}{k} \int_{-\infty}^{\infty} d\underset{\sim}{k}' \frac{e^{i(\underset{\sim}{k}'-\underset{\sim}{k}).\underset{\sim}{R}}}{k^2 k'^2} \tag{5}$$

We now proceed to prescribe a technique for the numerical integrations. First we separate the form factor $F(|\underset{\sim}{k}-\underset{\sim}{k}'|)$ into the radial and angular parts[7].

$$F(|\underset{\sim}{k}-\underset{\sim}{k}'|) = \sum_{\ell=o}^{\infty} F_\ell(k,k')P_\ell(\cos\theta) \tag{6}$$

in which

$$F_\ell(k,k') = \frac{2\ell+1}{2} \int_o^{\pi} F(|\underset{\sim}{k}-\underset{\sim}{k}'|)P_\ell(\cos\theta)\ \sin\theta d\theta \tag{7}$$

where $P_\ell(\cos\Theta)$ is the orthonormal set of the Legendre polynomials. The plane wave $e^{i\underset{\sim}{k}.\underset{\sim}{R}}$ may be expanded into partial waves,

$$e^{i\underset{\sim}{k}.\underset{\sim}{R}} = 4\pi \sum_{\ell=o}^{\infty} \sum_{m=-\ell}^{\ell} i^\ell j_\ell(kR)Y^*_{\ell m}(\hat{k})Y_{\ell m}(\hat{R}) \tag{8}$$

where $Y_{\ell m}$ are the spherical harmonics and j_ℓ the spherical Bessel functions. Substituting (6) and (8) into (2), followed by applications of the orthonormality condition of the spherical harmonics and the identity

$$\sum_{m=-\ell}^{\ell} |Y_{\ell m}(\hat{R})|^2 = \frac{2\ell+1}{4\pi} \tag{9}$$

we obtain

$$IFL = \frac{-4}{\pi^2} \int_o^\infty dk \int_o^\infty dk' \sum_{\ell=o}^\infty j_\ell(kR) j_\ell(k'R) F_\ell(k,k') F(k) F(k') \quad (10)$$

Instead of a previously six dimensional integral we now have an integral of only three dimensions but containing an infinite summation over ℓ. Fortunately the series converges rapidly and only a small number of ℓ is needed for convergence.

Except for the simplifying transformations of k and k' into k = $\frac{1}{2}(\tilde{K} + \tilde{T})$ and k' = $\frac{1}{2}(\tilde{T} - \tilde{K})$ in (3), the reductions of (3) and (4) are similar and straightforward. Thus

$$IFF = \int_o^\infty dk |F(K)|^2 K j_o(KR) \quad (11)$$

and

$$ILL = \left\{\frac{1}{2\pi} \int_o^\infty dk |F(k)|^2 j_o(kR)\right\}^2 \quad (12)$$

IFF and ILL are both one dimensional integrals whose solutions may be computed to any desired degree of accuracy, for instance, by the Simpson's six point integration. IFL, on the other hand, is an example of a three dimensional integral whose accurate evaluation requires a number of mest points and in Fortran language means an enormous amount of CPU time. This difficulty may be circumvented by first performing the integration of the function F (k,k') given by (7) for appropriately selected values of k and k' whose results amy be stored, say, in an array FSUBL. Explicitly we have

Function	Integration	Array
$F_0(k,k')$	$\frac{1}{2}\int_{-1}^{1} F(\sqrt{k^2+k'^2-2kk'x})dx$	FSUB0
$F_1(k,k')$	$\frac{3}{2}\int_{-1}^{1} F(\sqrt{k^2+k'^2-2kk'x})xdx$	FSUB1
$F_2(k,k')$	$\frac{5}{2}\int_{-1}^{1} F(\sqrt{k^2+k'^2-kk'x})\ \frac{(3x^2-1)}{2}\ dx$	FSUB2
⋮	⋮	⋮

This way the three dimensional integral is reduced to an equivalent pseudo two dimensional integral. The array FSUBL may be recalled subsequently and then efficiently interpolated for the necessary values of k and k' during the resulting double integration. The obvious advantage of this procedure comes about in not having to perform the same calculations $m \times n_{max}^3$ times ($\sim 10^{10}$) for each value of ℓ for each atom-atom pair where $m \sim 20$ is the number of internuclear distances and $n_{max} \sim 1000$ is the maximum number of mesh points required to reach convergence.

Except for the smaller number of significant figures which is solely a function of machine time, the values of IFL, IFF and ILL on H-H and He-He in the present calculations agree with the results obtained by analytic methods[8].

3. ACKNOWLEDGEMENTS

The author wishes to express his appreciation for the useful comments and support of Professor B. Linder and helpful discussions with Professor Chew-Lean Lee. This work was supported by the National Institute of Health, Grant No. GM23223.

REFERENCES

1. Linder, B. and Rabenold, D.A.: 1972, Adv. Qaunt. Chem. 6, p.203.

2. Linder, B., Lee, K.F., Malinowski, P. and Tanner, A.C.: 1980, Chem. Phys. 52, p.353.

3. Unsöld, A.: 1927, Z. Physik 43, p.563.

4. Clementi, E. and Roetti, C.: 1974, At. Data Nucl. Data Tables 14, p.177.

5. Malinowski, P., Tanner, A.C., Lee, K.F. and Linder, B.: in press, Chem. Phys..

6. Malinowski, P., Tanner, A.C., Lee, K.F. and Linder, B.: in press, Int. J. Quant. Chem..

7. Brink, D.M. and Satcher, G.R.: 1971, Angular Momentum, Clarendon Press, p.102.

8. Lee, K.F.: 1981, Ph.D. Dissertation, Florida State University, Tallahassee, FL U.S.A.

AUTHOR INDEX

SUBJECT INDEX